CPC UCW
LLYFRGELL
LIBRARY
ABERYSTWYTH

AF616256

The Political Economy of Environmental Protection

CONTEMPORARY STUDIES IN ECONOMIC AND FINANCIAL ANALYSIS VOL. 24

Editors: Professor Edward I. Altman and Ingo Walter, Associate Dean
Graduate School of Business Administration, New York University

CONTEMPORARY STUDIES IN ECONOMIC AND FINANCIAL ANALYSIS

An International Series of Monographs

Series Editors: Edward I. Altman and Ingo Walter
Graduate School of Business Administration, New York University

Volume 1. DYNAMICS OF FORECASTING FINANCIAL CYCLES: Theory, Technique and Implementation
Lacy H. Hunt, II, *Fidelcor Inc. and the Fidelity Bank*

Volume 2. COMPOSITE RESERVE ASSETS IN THE INTERNATIONAL MONETARY SYSTEM
Jacob S. Dreyer, *Graduate School of Arts and Sciences, New York University*

Volume 3. APPLICATION OF CLASSIFICATION TECHNIQUES IN BUSINESS, BANKING AND FINANCE
Edward I. Altman, *Graduate School of Business Administration, New York University,* Robert B. Avery, *Graduate School of Industrial Administration, Carnegie-Mellon University,* Robert A. Eisenbeis, *Federal Deposit Insurance Corporation,* and Joseph F. Sinkey, Jr., *College of Business Administration, University of Georgia*

Volume 4. PROBLEM AND FAILED INSTITUTIONS IN THE COMMERCIAL BANKING INDUSTRY
Joseph F. Sinkey, Jr., *College of Business Administration, University of Georgia*

Volume 5. A NEW LOOK AT PORTFOLIO MANAGEMENT
David M. Ahlers, *Graduate School of Business and Public Administration, Cornell University*

Volume 6. MULTINATIONAL ELECTRONIC COMPANIES AND NATIONAL ECONOMIC POLICIES
Edmond Sciberras, *Science Policy Research Unit, University of Sussex*

Volume 7. VENEZUELAN ECONOMIC DEVELOPMENT: A Politico-Economic Analysis
Loring Allen, *University of Missouri, St. Louis*

Volume 8. ENVIRONMENT, PLANNING AND THE MULTINATIONAL CORPORATION
Thomas N. Gladwin, *Centre d'Etudes Industrielles and New York University*

Volume 9. FOREIGN DIRECT INVESTMENT, INDUSTRIALIZATION AND SOCIAL CHANGE
Stephen J. Kobrin, *Sloan School of Management, Massachusetts Institute of Technology*

Volume 10. IMPLICATIONS OF REGULATION ON BANK EXPANSION: A Simulation Analysis
George S. Oldfield, Jr., *Graduate School of Business and Public Administration, Cornell University*

Volume 11. IMPORT SUBSTITUTION, TRADE AND DEVELOPMENT
Jaleel Ahmad, *Concordia University, Sir George Williams Campus*

Volume 12. CORPORATE GROWTH AND COMMON STOCK RISK
David R. Fewings, *McGill University*

Volume 13. CAPITAL MARKET EQUILIBRIUM AND CORPORATE FINANCIAL DECISIONS
Richard C. Stapleton, *Manchester Business School and New York University*, and M. G. Subrahmanyam, *Indian Institute of Management and New York University*

Volume 14. IMPENDING CHANGES FOR SECURITIES MARKETS: What Role for the Exchanges?
Ernest Bloch and Robert A. Schwartz, *Graduate School of Business Administration, New York University*

Volume 15. INDUSTRIAL INNOVATION AND INTERNATIONAL TRADING PERFORMANCE
William B. Walker, *Science Policy Research Unit, University of Sussex*

Volume 16. FINANCIAL POLICY, INFLATION AND ECONOMIC DEVELOPMENT: The Mexican Experience
John K. Thompson, *Mellon Bank, N.A.*

Volume 17. THE MANAGEMENT OF CORPORATE LIQUIDITY
Michel Levasseur, *Centre d'Enseignement Superieur des Affaires, Jouy-en-Josas, France*

Volume 18. STABILIZATION OF INTERNATIONAL COMMODITY MARKETS
Paul Hallwood, *University of Aberdeen*

Volume 19. PROBLEMS IN EUROPEAN CORPORATE FINANCE: Text and Cases
Michel Schlosser, *Centre d'Enseignement Superieur des Affaires, Jouy-en-Josas, France*

Volume 20. THE MULTINATIONAL ENTERPRISE: International Investment and Host-Country Impacts
Thomas G. Parry, *School of Economics, University of New South Wales*

Volume 21. POLICY RESPONSES TO RESOURCE DEPLETION: The Case of Mercury
Nigel Roxburgh, *Management Studies Division, The Economist Intelligence Unit Limited*

Volume 22. THE INTERNATIONAL MONEY MARKET: An Assessment of Forecasting Techniques and Market Efficiency
Richard M. Levich, *Graduate School of Business Administration, New York University*

Volume 23. TRANSNATIONAL CONGLOMERATES AND THE ECONOMICS OF DEPENDENT DEVELOPMENT: A Case Study of the International Electrical Oligopoly and Brazil's Electrical Industry
Richard Newfarmer, *University of Notre Dame*

Volume 24. THE POLITICAL ECONOMY OF ENVIRONMENTAL PROTECTION
Horst Siebert, *University of Mannheim*, and Ariane Berthoin Antal, *International Institute for Environment and Society, Science Center Berlin*

The Political Economy of Environmental Protection

by HORST SIEBERT
Lehrstuhl für Volkswirtschafts-Lehre und Aussenwirtschaft
Universität Mannheim

ARIANE BERTHOIN ANTAL
International Institute for Environment and Society,
Science Center Berlin

JAI PRESS INC.
Greenwich, Connecticut

Library of Congress Cataloging in Publication Data

Siebert, Horst, 1938–
The political economy of environmental protection.

(Contemporary studies in economic and financial analysis; v. 24)
Bibliography: p.
Includes index.
1. Pollution—Economic aspects. 2. Environmental protection—Economic aspects. 3. Environmental policy—Economic aspects. I. Antal, Ariane Berthoin, joint author. II. Title. III. Series.
HC79.P55S55 614.7 78-13843
ISBN 0-89232-116-4

165 West Putnam Avenue
Greenwich, Connecticut 06830

ISBN NUMBER: 0-89232-116-4
Library of Congress Catalog Card Number: 78-13843
Manufactured in the United States of America

Contents

Preface

The natural environment and its relationship to the economic system are the subject of this book. It is a bitter fact that air and water, beautiful landscapes and raw materials, minerals and energy reserves—in short, the resources of nature—have fallen from the paradise of free goods to the realm of scarcity. These goods have become scarce both as public consumer goods and as factors of production. The land of milk and honey, the land of superabundance, has disappeared and the environment's complex interdependencies and natural regulations, which society ignored for so long, must now be learned.

This book was originally published in German under the title *Das produzierte Chaos—Oekonomie und Umwelt.* We are grateful to the German publishers for releasing the material. The book has been completely rewritten with an international frame of reference, and the examples originally oriented to the German reader have been revised for the international audience.

A book grows out of many impulses and stimuli. Some of these stem from ideas, some from individuals. We would like to thank Blair T. Bower, Ron Cummings, Ralph d'Arge, Robert Hawkins, I. Krutilla, Toby Page, Cliff Russell, William Schultze, and Walter Spofford, who made it possible for H. Siebert to become familiar with problems of environmental economics during a research leave. They also provided valuable insights for the discussion. Special thanks are offered to Allen Kneese and to Ingo Walter for constant encouragement. H. Siebert is grateful, too, to the students at the University of Mannheim for their constructive critical comments in courses. H. Bender suggested some examples of ecological damage. Acknowledgments are also due the members of the staff at the International Institute for Environment and

Society at the Science Center Berlin for their encouragement and support. The authors would especially like to thank David Antal, whose perceptive editorial work was an invaluable contribution to preparing the English manuscript. May the reader view what errors remain through the wisdom of Nietzsche: "The fallibility of a theory is not the least of its appeals; precisely therein lies its attraction for better minds."

Chapter I

Five Weeks, Five Days, Five Minutes

Ye who love the haunts of Nature,
Love the sunshine of the meadow,
Love the shadow of the forest,
Love the wind among the branches,
And the rain-shower and the snowstorm
And the rushing of great rivers
Through their palisades of pinetrees,
And the thunder in the mountains,
Whose innumerable echoes
Flap like eagles in their eyries . . . .

—Henry Wadsworth Longellow
The Song of Hiawatha

The Borinage, Belgium, 1930: During five foggy days sixty people died as a result of a sulfur dioxide concentration in the air ten times greater than in the same week of previous years. 6,000 people fell ill.

Donora, Pennsylvania, October 1948: The gravedigger prepared twenty graves instead of the usual two, and 42.8 percent of the population of 6,000 was stricken in five days by poisonous industrial emissions which stifled the city and threatened all forms of life during an atmospheric inversion.

London, December 1952: During a week-long heavy fog loaded with carbon monoxide and other toxic substances, 4,000 more deaths occurred than in previous Decembers.

Are these freak examples? What about the world-reknowned case of hydrogen sulfide emissions in Poca Rica, Mexico? And the Tokyo-Yokohama asthma, a new discomfort experienced regularly enough to be classified as a sickness? And the increased mortality rate in the Ruhr industrial area and the environs of Cologne, which scientists attribute to air pollution?

What of the smog of Los Angeles? There, in 1971, toxic substances in the air exceeded medically established standards 272 days of the year. In other words, two thirds of the year, or three of the four seasons, were characterized by unsafe conditions. The ozone layer of the stratosphere protects the earth from deadly ultraviolet rays, but we are letting nitrogen oxides[1] and Freon from aerosol cans destroy it.[2]

Acidic rains carrying the emissions of industrial areas in the German Ruhr valley and in England are falling on the Swedish countryside. In the region between New York and Maine the pH value of rain has been sinking, indicating that the rain is becoming more acidic there, too.[3] This phenomenon appears to be affecting the entire northern hemisphere.

Are flukes of nature responsible for the contamination of our rivers by phosphorus, DDT, and mercury? Statistical analysis suggests that there are significant correlations between illnesses and environmental quality.[4] Many of the wastes which are dumped into the environment, such as lead, cadmium, and vinylchloride, are suspected of being dangerous, even carcinogenic.

A similarly disquieting research result is that the milk of American mothers contains a higher concentration of DDT than is permitted by the U.S. Food and Drug Administration in marketed cow's milk. German children are also fed this unhealthy mixture. "Significant proportions" of DDT were detected in the milk of 136 out of 137 carefully chosen mothers. Only one of the women produced milk which met the standards set on food products.[5]

Can we still be dealing with exceptional cases when densely populated centers such as Tokyo, New York, and the Ruhr experience high levels of pollution frequently enough to cause alerts? Can we dismiss the sound of air-raid sirens—now used as smog alarms—as an exception? How far have we allowed pollution to go when the American Medical Association (AMA) considers it necessary to publish the *Physician's Guide to Air Pollution Episodes* as a handbook for treating patients in case of a yellow or red alarm? Or when doctors recommend that patients suffering from respiratory, heart, or circulatory ailments confine themselves to bed? Or that patients install special air filters alongside the air conditioners in their homes? Or that they leave the area temporarily, or even permanently?

Our environment is so degraded that we can no longer afford to wait for a solution. But what is the environment? It is more than the flora and fauna on the endangered species lists, like the golden frog arrow of the orchid family and the Indiana bat. The environment is more than an idyllic spot where cuckoos call and foxes bark. The environment is our most basic living space—the air we breathe, the water we drink, the space we live in. It includes—and now for that cuckoo's call—the delights and relaxation of nature.

In the past, people have turned to heaven to plead for scarce basic goods, such as bread. Now the environment itself—pure water and clean air—have fallen into this category. Is it time to revise the Lord's Prayer?

Besides praying, the scientist must analyze the problem, develop solutions, and recommend possible programs. Such is the goal of this book. But what, one might ask, can an economic perspective contribute to the solution of environmental problems? Some might contend that the environment poses much too important a problem to leave to economists. The question is a meterological one, one of water management and hydrology. Environmental pollution is a physical, a chemical—in short, a natural science problem. In addition, the environment has legal, medical, social, and political implications. Environmental pollution involves all of these, and each discipline contributes solutions. But environmental pollution is *also* an economic problem. The economic activities of production and consumption are significant causes of pollution, and, therefore, the solutions will also have to be sought in economics.

Five weeks without food, five days without water, five minutes without air are the outer limits of human tolerance. In spite of this, we treat these crucial goods as though they were unimportant, as though they did not exist and did not need to exist. They are not taken into account by economic theories and decision-making models. For economists, air and water have long been the prototypes of free goods, available in unlimited quantities, for all purposes, at zero price. And they are still considered free goods for commercial activities, in which they are most welcome as free receptacles for all kinds of wastes.

Five weeks without food—five days without water. Should not these biological limits of humans be reflected in values assigned to goods produced and consumed? Can such elementary goods as water and air continue to be assessed at zero price while trifles like the chrome trimmings on the latest automobile model and the automatic garage door opener rise in price and value? Can we continue to exploit the environment for all competing purposes? Can we at the same time drink water and dump into it industrial wastes like salts, acids, ammonium chlorides, DDT, nitrates and phosphates, and the extravagantly applied fertilizers that are later washed out from the soil?

Five weeks without food—five minutes without air. Societal well-being continues to be measured only according to the goods produced within a given period. But this gauge, called the Gross National Product (GNP), is invalid when it fails to take into account pollutants and when it continues to rise from year to year while the environment deteriorates. Environmental quality is also an element of societal well-being. The degradation of our air, the contamination of our water, the disfigurement of our landscapes, and the destruction of nature's charms should be included into our yearly balance sheets.

Five days without water—five minutes without air. The quality of life has assumed new value, and the environment has been accorded greater significance as a determining factor. Our respect for the environment has grown alongside our appreciation of its functions and discoveries about its five-billion-year history. We have gradually come to understand that the system is extremely delicate and complex. We cannot repair damages we might incur to the earth's oxygen layer. We have had to recognize that, in spite of all our scientific knowledge, we can only most inadequately describe the extraordinarily intricate and manifold interactions of the laws of nature. Any attempts to predict possible consequences of our actions are sadly unreliable.

This book deals with the environment from the perspective of the economist. It examines the degree of environmental degradation and its causes. It discusses the relationship between the economy and the environment, between the benefits and costs of environmental protection, and it examines necessary changes in the economic system, potential solutions, possible measures, and probable effects of these measures. We do not, however, wish to address economists only. The environmental problem has expanded beyond the confines of scientific journals and specialists' lectures to become part of the daily concerns of individuals around the world. It must therefore be discussed in practical rather than purely abstract or technical terms. Such has been our intent.

NOTES

1. See the results of the study by the Dutch meteorologists, Paul Crutzen, in *Time,* Feb. 23, 1976.

2. *Bild der Wissenschaft,* February 1975.

3. *Chemical and Engineering News* 23 (November 1976).

4. T. Page, R. H. Harris, and S. S. Epstein, "Drinking Water and Cancer Mortality in Louisiana," *Science* 193:55–57 (1976).

5. *Der Spiegel* 8, 1975.

Chapter II

Pollutants in Millions of Tons: Some Empirical Findings

And the fish that was in the river died; and the river stank, and the Egyptians could not drink of the water of the river. . . .

Exodus 7:21

Anyone who has driven through Los Angeles on a hot summer day has noticed the constant cloud of fumes. The physical effect is blandly referred to as "slightly irritating." In Los Angeles county alone 9,100 tons of carbon monoxide are spewed into the atmosphere each day, and the photochemical smog bank reaches over 100 miles eastward. The smog report, which lists the observed concentrations of the major pollutants in the air and predicts whether your eyes will water, has become an integral part of the Los Angeles *Times* daily weather report.

The air we breathe today is no longer the good old mixture of 78 percent nitrogen, 21 percent oxygen, 0.95 percent argon, 0.03 percent carbon dioxide, and small amounts of several other pure gases. A number of pollutants are now floating around which our bodies absorb through our lungs and skin. The stone saints of the Cologne Cathedral have been eaten away by this mixture; what of the delicate tissues of our lungs?

CARBON DIOXIDE-FREE LUNGS

In 1970 in the United States 100 million tons of carbon monoxide, 33.3 million tons of sulfur dioxide, 22.1 million tons of oxides of nitrogen, 23 million tons of hydrocarbons, and 22.3 million tons of particulates were produced.[1] The following quantities were estimated for West Germany in 1972: (in millions of tons) CO_2 = 6.8; SO_2 = 3.5; NO_2 = 1.7–2.1; HC = 0.6–0.8; and particulates = 0.8.[2] Table 1 summarizes the emission trends of air pollutants in the United States from 1940 to 1970.

Table 1.[3] Emission trends of air pollutants in the United States 1940—1970 (in millions of tons)

Year	CO	SO_2	NO_2	HC	Particulates
1940	85	22	7	18	27
1950	103	24	10	26	26
1960	128	23	14	32	25
1968	150	31	21	35	26
1969	154	34	22	35	27
1970	147	34	23	35	25

Carbon monoxide (CO) is a colorless, odorless, poisonous gas, more than two thirds of which is produced by the incomplete combustion of gasoline in cars. Since carbon monoxide has a strong affinity to hemoglobin, which carries oxygen in the blood, it is transported into the cells and thus impairs the oxygen supply. Headaches and slowed reaction result. People suffering from heart, lung, and blood vessel disorders are the most severely affected.

Sulfur dioxide (SO_2), and to some extent *sulfur trioxide* (SO_3), is a corrosive, poisonous gas which develops from the burning of sulfurous fossil fuels, such as coal and oil. A single cubic meter of heating oil contains 20–40 kilos of sulfur. It is estimated that of the 32.6 percent of SO_2 which fell on the United States in 1971, 90 percent can be attributed to furnaces, 6 percent to iron-ore pelleting, 2.5 percent to automobile exhausts, and 1.5 percent to sulfuric acid production.[4] According to statistics from the German Federal Ministry of the Interior,[5] sulfur dioxide concentration is higher in many urban areas than the long-term maximal levels of 0.4 mg/m^3 set by the air pollution bureau, causing irritation to eyes and the upper respiratory tract. Attached to fine dust, SO_2 also enters the lungs and destroys the delicate tissue. Sulfuric acid, which forms when sulfur trioxide mixes with moisture, has the same effect. Other effects of SO_2 are the acidic rains which fall both in industrial and in rural areas. The last chemically neutral rains in Frankfurt were recorded in the early 1950s.

Oxides of nitrogen (NO and NO_2) are produced in combustion processes at high temperatures. Out of the 22.7 million tons in the United States, 10 million tons come from furnaces, 11.7 million tons from automobiles, and 1 million tons from other sources, such as chemical plants. The side effects on humans are similar to those experienced from sulfur dioxide: eye and respiratory tract irritation, respiratory difficulties, dizziness, and temporary paralysis. A concentration of 250 parts per million (ppm) is lethal; the official limit of 0.5 ppm is exceeded in many cities.

Hydrocarbons (HC) are also the result of incomplete combustion processes in automobiles and furnaces, with oil refineries and petrochemical plants being other major sources. Some hydrocarbons have been found to produce tumors in animals, and researchers believe that one of them, 3.4-benzyprene, could be one of the chemical substances causing cancer. Surveys taken since 1962 indicate that concentrations of 3.4-benzyprene have increased.

Particulates, or dusts are generated from the combustion of solid fuels and from extraction, processing, and use of all minerals. Coarse particles, such as soot, which caused serious problems in the past, have been reduced, but fine particulates have multiplied significantly. In addition to common household dust and ashes, the atmosphere now contains many other ingredients, such as lead (which can attain concentrations of 0.02 mg/m^3 in West German industrial centers under unfavorable conditions), iron, copper, zinc, and small amounts of nickel, cadmium, vanadium, cobalt, and chrome. Other elements which are bonded to dust particles, such as DDT and the carcinogenic hydrocarbons mentioned above, also come under this category. The metal particles in the air have an especially destructive effect on cells.

These five pollutants—carbon monoxide, sulfur dioxide, oxides of nitrogen, hydrocarbons, and particulates—are the best-known contributors to air pollution; however, they are only the tip of the iceberg. The lists of other components are not yet complete. Some scientists identify over 300 pollutants, while others claim a minimum of 500. Environmental analysis is in its infancy and many pollutants are still unknown. The interrelationships between individual chemical elements, the human organism, and the various environmental media are extraordinarily complex. Extensive research is necessary before many substances can be identified as pollutants. The best example is DDT, the discovery of which earned the Swiss chemist Dr. Paul Mueller the Nobel prize for medicine in 1948. The use of this chemical is now forbidden in most countries.

A number of other pollutants should not be forgotten:

—Biocides, that is, chemicals used to destroy insects, weeds, and fungi. One common biocide is PCB (Polychlorinated biphenyls) the side effects of which are similar to those of DDT. Worn tires release PCB throughout the world.

—Hydrochlorides, which are released in the combustion of chloric hydrocarbons (such as synthetic products) in waste incineration.

—Fluoride compounds, which are produced in energy plants and in aluminum, iron, steel, and stone industries. They have been proven to cause plant damage.

—Gases and vapors, such as inorganic sulfur compounds (hydrogen sulfide, carbon disulfide) and inorganic compounds such as phenols, amines, nitrogen compounds, mercaptan, aldehydes, etc. Sources of these pollutants are oil manufacturing and food industries, automobile exhausts, and commerical animal farms.

The presentation of empirical findings in terms of tons can be misleading. When dealing with pollutants, one ton is not simply a ton, and the effects on the environment vary with the pollutant. Fluoride compounds, for example, are a hundred times more destructive to plants than an equivalent amount of sulfur dioxide emissions. And one ton of cadmium is much more devastating than a ton of oxides of nitrogen. In addition, pollutants have the unpleasant characteristic of interacting with each other to produce new pollutants. This combining of different pollutants, a phenomenon referred to as synergism, can have a more powerful effect than do the elements individually. Gaseous sulfur oxides and oxides of nitrogen, for example, react with cation particles to develop sufates and nitrates. Atmospheric synergism also causes the formation of acids from sulfur dioxide, producing acid rains. A most familiar result of such interaction of pollutants in the air is photochemical smog, a mixture of gases and particulates which is produced under strong ultraviolet rays, mainly from oxides of nitrogen and unburned hydrocarbons. This process results in ozone, peroxyacetyl (PAN), and minute, presumably liquid, particulates smaller than 1 micron (1 micron = 1/1000 millimeter) generating a particularly unpleasant type of fog.

Our air today is a mixture which comprises the above-mentioned pollutants and a great many unidentified foreign elements in addition to the natural ingredients of "fresh air," a rare phenomenon in cities. The concentrations in which these pollutants are found depend on local industries, furnaces and combusted materials, the volume of traffic, and the environmental conditions, such as sunlight, inversion layers, and fog. The Los Angeles type of smog in the warmer zones has its equivalent in the London smog of the middle latitudes, characterized by the occurrence of sulfur dioxide together with fine particulates and other pollutants which are intensified by inversion layers and natural fog.

FLUSH THE TOILET, THEY NEED THE WATER IN ST. LOUIS

Not only is the air we breathe saturated with pollutants from our economic development; our water is also contaminated. Germany's celebrated Rhine transports 24 million tons of poisons and pollutants to the Dutch frontier yearly. Daily, this represents about 3 tons of arsenic, 450 kg of mercury, 30,000 tons of sodium chloride ions, 16,000 tons of sulfates, 2,200 tons of nitrates, and over 100 tons of phosphates. Every day! Twenty-five million people draw their drinking water from this river.

The rich reserves of the seas could help us obtain the necessary foodstuffs to stave off hunger in the world. Yet into the oceans we dump our household, shipping, and industrial wastes, polluting the waters with oil spills, acids, poisons, and solid wastes, because no other receptacle or method of disposal is so cheap. In 1971, 29 percent of U.S. streams and shoreline miles were declared polluted,[6] and some beaches are not fit for bathing—quite an achievement for the so-called cultured societies of highly developed countries. Four million tons of untreated sewage are dumped into the waters of the New York coast each year. A type of "Dead Sea" has developed in which neither plants nor animals have a hope of survival. The bacterial content multiplied ten times over in the 1960s. If a fish swims into this water, its scales rot off. The planners of this sewage disposal project thought that the site was far enough from the shore, but the dead zone is moving closer to the coast.[7]

Our lakes are also suffering from this harsh treatment. The normal eutrophication, or aging, process has been accelerated by pollution.[8] Eutrophication is caused by the superabundant production of algae as a result of increased phosphate levels in the water. The algae become so thick that sunlight cannot penetrate to the lower levels of the sea. The flora die and their decomposition requires oxygen, thereby reducing the oxygen content of the water and suffocating all forms of life in the body of water. The death of Lake Erie is a world-famous example. The same problem is being faced in other countries as well. Lakes Geneva, Constance, and Bourget, for example, are also endangered.

Water quality is usually measured according to its capacity to sustain animal and plant life. This capacity is influenced by the oxygen content, the presence of foreign elements, and by the temperature of the water. The three factors are interdependent. For example, the presence of organic and inorganic foreign elements decreases the oxygen level of the water. The oxygen content also falls when the water temperature rises. Many other factors must also be taken into consideration, such as the depth, flow rate, and atmospheric conditions. The level of oxygen necessary to sustain life in fresh water is 10 mg per liter, and when the oxygen content falls below 2 mg, the body of water is considered biologically dead.

The oxygen in water is produced by the photosynthesis of aquatic flora and of phytoplankton. This oxygen is used in the bacterial decomposition of dead algae, and is therefore indispensable to the self-purification processes in the water. The amount of oxygen required by the bacteria to decompose organic matter is referred to as the biochemical oxygen demand (BOD). This concept is usually expressed in terms of the milligrams per liter of water used for decomposition in five days, and is, therefore, a measure of the amount of organic waste in the water. If there is not enough oxygen to decompose the wastes, decomposition occurs through anaerobic processes which produce foul-smelling sulfur oxides and methane gas. The water turns a dark color; fish die off.

Unfortunately, the foreign matter present in water cannot be measured by such a uniform scale as the BOD. Included in this class of pollutants are dissolved and undissolved matter; toxins such as mercury and cadmium; pesticides; acids; radioactive particulates; inorganic microorganisms; detergents and phenols, to list but a few. In contrast to organic pollutants, these elements are not biodegradable. They find their way into the sea or the ground water, or they can be stored or transported. Whereas scientists estimate the types of pollutants present in the air to range between 300 and 500, none even guess at the number of foreign elements in our waters. "Their number is theoretically unlimited and is steadily increasing,"[9] states an FRG official report. Further, "the chemical industries produce about 500 new compounds each year which later turn up as pollutants in water and have to be traced."[10]

Most foreign elements in water are attributable to untreated or inadequately treated industrial effluents which are dumped into the water. In second place is agriculture—in spite of the claims of farmers that they are providing society with free landscape conservation, on the basis of which they will certainly demand new subsidies. Runoffs of the increasingly employed chemical fertilizers from agricultural land have led to significant accumulations of phosphates and nitrates in water. Biocides, insecticides, and herbicides have seriously degraded our water. Many of these chemicals decompose only very slowly (decades!) and thus some, such as DDT and polychloride biphenyls have spread around the world.

Traces of DDT are found everywhere: in Antarctic penguins, in human mothers' milk, in oxygen-producing plankton of the oceans, and the now extinct Louisiana pelicans. Most importantly, concentrations of DDT and similar chemicals in food chains amplify a millionfold the DDT content of individual organisms. This accumulation process has been proven conclusively[11]: the water tested in Long Island Sound contained 0.000003 ppm DDT. The zooplankton, whose fatty tissues retain DDT, already experienced 1000 times higher a concentration (0.04 ppm). In small fishes from the same water 0.5 ppm were found—a tenfold increase; in large fishes, 2 ppm; and in fish-eating cormorants the concentrations spiraled to 25 ppm. In other words, the DDT content was amplified *one million times* just within this short food chain.

The third major polluter, in addition to industry and agriculture, is the household. Certain products employed in the home, such as detergents and phosphates, also contribute to water degradation. Pollutants also fall into water from the atmosphere by precipitation, like dust deposits and rain. On the other hand, pollutants can also partially return to the atmosphere when they evaporate from the sea.

The capacity of water to sustain plant and animal life, and the absence of dead fish are one measure of water quality. But it is an insufficient standard for human purposes. Substances which roach and whitefish can tolerate might well be harmful to humans. Humans use water for many purposes—for drinking, irrigation, sports, and in industrial processes. The definition of pollution levels should therefore depend on the intended use and on the purity required for that use. Society must decide which of these uses must be accorded priority and it must also define the criteria according to which the water quality can be measured, in addition to determining which elements constitute pollutants. The next step is to develop the know-how and the equipment to analyze the water for all these pollutants which, even in the smallest concentrations, can cause health damage through continuous accumulation.

THE CINDERELLA OF ECONOMICS

As far as economists were concerned, the environment did not use to exist. It was a free good which carried no price tag and was available in unlimited quantities. There was no need to economize on it. And economists do not tread where there are no goods to be allocated. But the environment is no longer a free good; it must be rationed. What role does this scarce resource now play for the economist? What are the relationships between the environment and the economy?

The environment provides the economy with inputs into a wide variety of production and consumption processes. For example, the consumer activity of swimming comprises various inputs from the environment, such as the temperature and turbidity of the water, pollutant content, biochemical oxygen demand, location of the river, etc. Similarly, the consumption of air involves a specific mixture of inputs, such as the basic hydrogen, oxygen, argon, carbon monoxide, and our pollutant additives. The beauty and peacefulness of a landscape contribute to our enjoyment of leisure hours. All these elements are inputs into our existence, and are therefore included in the broad concept of environmental quality. Let us now look at the economic aspects of these environmental components of our lives; we will draw our illustrations primarily from the environmental media of air and water, but the concepts also apply to the other media.

In the language of the economist, the environment is a *public good* and as such is characterized by two criteria:

—A public good, as opposed to a private good, can be used by many individuals. Collective use is not enough, however, to distinguish a public good, inasmuch as certain private goods also involve collective consumption (what would a football game be without the shouts of the crowd?).

—A public good, technically, does not permit the exclusion of other competing users. The classic example is the lighthouse, which can be used as an orientation point by all vessels, even if an individual fisherman does not expect to pay for its upkeep. This technical impossibility of exclusion is, in many cases, a characteristic of a public good. But we should not underestimate human ingenuity when it comes to getting around the technical problems of excluding others from consuming collective goods. As mentioned above, public events are collective goods, yet measures to exclude people from them have been devised. From fences to television cables, from toll booths on roads to electricity meters, people have not lacked in imagination in designing methods to circumvent the technical difficulties of exclusion. Therefore, we must define a public good as one which can be used by all and from which no one could, or should, be excluded. This *should* is the deciding factor, so the nonexclusion criterion is often a matter of value judgment.

Environmental services as consumer goods are a public good according to this definition. The good can be used by all, and technical exclusion, if it were possible, is socially and politically undesirable. The reader who has doubts or questions at this point about the definition, and specifically about the two criteria identified above, should keep them in mind. We will return to a more detailed examination of this concept, and will make significant adjustments in the determining characteristics of public goods later.[12]

In addition to providing public consumer goods, the environment furnishes the economic system with resources and raw materials which are used as inputs into production activities: energy, minerals, metals, etc. Resources are used to produce goods which, in turn, are directed into consumption. Arrows 1 and 2 in Diagram 1 below illustrate these relationships between the economic system and the environment. The economic system comprises the raw materials and the pollutants, as well as inputs into consumption.

Wastes are inevitable by-products of both production and consumption activities. By-products are generated, for example, when several goods are manufactured in one process. Many by-products are of

Diagram 1 Interactions between the environment and the economic system

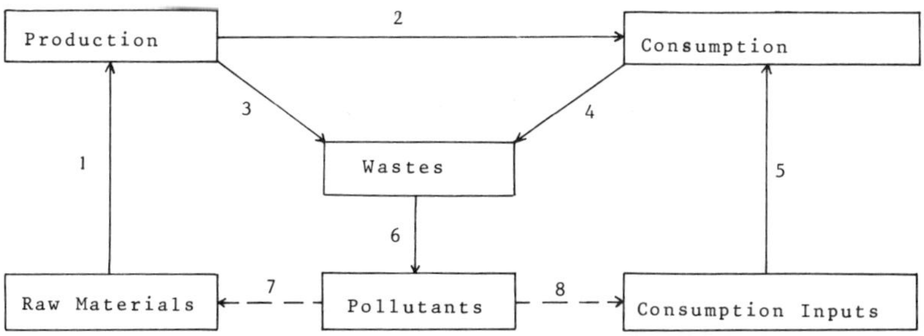

economic value, such as heating oil and gasoline which are produced simultaneously in the process of refining crude oil. Another example is coke and gas, which are produced together at coking plants. Often, however, no use can be made of the by-products: water used for cooling purposes in electric power plants, ashes from incineration, carbon monoxide released by our cars, or phosphates from our washing machines. Some of the by-products are useful in small amounts, such as some acids. But the exorbitant quantities in which they are manufactured cannot be used. An interesting study conducted on the chemical industry in the FRG revealed that this sector produced 48 million tons of goods in 1970, and their wastes totaled 12 million tons. In other words, the waste by-products represented one quarter of the output.[13]

Unwanted and unusable by-products of production and consumption are disposed of in the environment. This is the third role of the environment in the economic process. In addition to providing consumer goods and furnishing resources for production, the environment serves as a dumping ground for waste products. Since these by-products are unwanted, they must be disposed of somehow, and the environment is the cheapest way—from the short term and individual point of view.

The three functions of the environment in the economy are illustrated by arrows 1, 3, 4, and 5 in Diagram 1.

The environment can be conceived of as a capital good which at all times allows many types of uses. Through biological, physical and chemical processes this capital good can regenerate itself: new trees grow, fish populations increase again after big catches, organically polluted water can purify itself. These reproductive processes in the environment are characterized by two traits. First, they can only occur within certain limits. If too many trees are felled in a forest, the climate will change, the earth will become chalky, and the reproductive process will thus be brought to a halt. If water is too heavily polluted, its self-purification capacity is destroyed. A second characteristic of the environment as a capital good is that although the system can regenerate itself, humans cannot easily replace what they destroy, nor can they always correct the processes of change they have set in motion.

At this point in the discussion a major distinction must be drawn between waste products emitted, pollutants, and ambient quality. These are all key terms in environmental policy. Wastes are the unwanted by-products of production and consumption which are then emitted into the environment. Pollutants are the wastes present in a particular environmental medium at a specific time. Wastes become pollutants when they are diffused, accumulated, or structurally changed in the environment. The pollutant content of the medium determines its ambient quality. Arrow 6 in Diagram 1 describes the relationship between wastes and pollutants as a function of diffusion and transformation.

The distinction between wastes and pollutants is extremely important. The desired standard of environmental quality must be expressed in terms of the pollutants present in the medium, or its ambient quality. However, to prescribe intervention, environmental degradation must be discussed in terms of wastes emitted. This difference gives rise to a great ambiguity in economic policies, to which we will return later.

The pollutants which the environment cannot absorb or decompose affect the quality of environmental services. This is because the pollutants influence the characteristics of the inputs into the consumption process. As we saw earlier, consumption activities are defined by many traits, of which the presence or absence of pollutants is one. Pollutants can also affect production activities inasmuch as they deteriorate the quality of the inputs and thus make preliminary treatment necessary before use. There is, then, a link between pollutants and the services and resources provided by the environment. This function describes the effect on the consumer of changes in the quality of public consumer goods and the additional costs of the enterprise as a result of the change in the quality of inputs to production. It is referred to as the damage function (arrows 7 and 8). The determination of this damage function, which implies knowledge of the diffusion function, is the decisive precondition for the formulation of economic policy measures.

THE LAW OF CONSERVATION OF MATTER

Matter, customarily measured in terms of weight, can neither be created nor destroyed. This law of the conservation of matter[14] also holds true for the mass which is extracted from the environment in the form of raw materials. These materials are certainly transformed in the production and consumption processes—as the Latin roots *pro-cedere* indicate—a new good is made from a combination of already existing matter. The mass which is extracted from the environment is also returned to it. Expressed in physical terms, for example tons, the environment always receives back the same amount of matter as was initially removed in the form of raw materials. When coal burns, the mass of all the resulting products of combustion, such as flyash, solid ash, and gases equals the original mass, plus the oxygen used in the burning process. The total wastes delivered into the environment in any one economic period must roughly equal the weight of the combusted matter, the other resources used in the production process, and the foods and oxygen from the atmosphere used during that period. The economy is then interpreted as a through-put system, which receives material flows from the environment and returns approximately the same amount of material flows back to the environment.

The restitution can be delayed; it can occur in subsequent economic periods. This happens in three cases: capital goods, durable consumer goods, and recycling.

Capital goods have a longer life than one economic period. They are used over many periods. When they have depreciated, become obsolete, or are no longer usable, they are scrapped. Not until this point is the matter contained in capital goods returned to the environment.

Durable goods undergo a similar process. Goods such as cars and washing machines are not consumed in a single act, but are returned to the environment after several periods of use. Therefore, the concept of destruction or disappearance of matter associated with consumption must be understood as an activity in which benefits created grow out of the combination of many outputs.

Capital and durable consumer goods do not revise the identity between matter received from and returned to the environment during a period when the mass of the newly produced durable goods exactly equals the mass of discarded goods. On the other hand, if in each period more durable goods are manufactured than are scrapped, then a part of the matter remains in the economy rather than returning to the environment.

The reuse of material, recycling, is yet another possibility of retaining material in the economy for a longer period. Both discarded matter and by-products of production processes can be treated in this way. The relationship between the matter extracted from the environment and the wastes returned to it is illustrated in Diagram 1. The matter represented by arrows 1 and 5 equals that represented by arrows 3 and 4. In other words, the flows between the economy and the environment must be of identical weight.

The law of conservation of matter refers to the weight of all the goods drawn from and restored to the environment. This global approach is then broken down into the flows of individual materials and substances between the environment and the economy. It has, therefore, been suggested that the flows of the individual substances, their transformations in the economic process, and their return to the environment be portrayed in models, such as flow diagrams. Flow diagrams can also provide valuable information when read "backwards." By tracing the development of a pollutant back through various production activities, its source can be identified. The physical and chemical processes which generate it can be analyzed. An examination of the flows can indicate the point at which a substance entered the environment and can pinpoint its origin in the environment. Such flow diagrams are bound to be extremely complex. They must comprehend a broad range of chemical and physical processes in production activities. Flow diagrams discover the potential key points in the development chain of a pollutant at which technical corrective measures could be undertaken.

A POLLUTANT BALANCE SHEET

Flow diagrams are one way to illustrate the relationships between the environment and the economy. Another related concept is the balance sheet of pollutants or materials. Table 2 presents such an account. It lists for one area and one environmental medium (on the left) the source of the pollutants according to the original amount of a pollutant, the newly discharged pollutants, and those received from another medium or region. On the right side, the residue of pollutants is accounted for in terms of the pollutants assimilated by the environment, those disposed of elsewhere, the outlets into another medium or a different region, and lastly, the final stock of pollutants.

A system of such accounts would permit the description of changes in pollutants over a certain time span. Just as the economic activities of a

Table 2. Pollutant Account

Source of Pollutants	*Residue of Pollutants*
1. Initial amount	5. Assimilated pollutants
2. Access from another environmental medium	6. Pollutants removed
	7. Outlets into another medium
3. Access from another region	8. Outlets into another region
4. Pollutants produced in the region	9. Final stock of pollutants

nation can be accounted for ex post facto, so could an ecological account statistically report changes which have taken place in the environment.

In order to establish such a system, accounts should be drawn up for the major pollutants in each of the basic environmental media. Especially important is the development of a chart listing the pollutants differentiated by period, source, and eventual mode of disposal, again specified by region and environmental medium.

The pollutants to be included depend on the extent of damage each causes. Very little information is available in this respect, so a precondition for the implementation of such a system is the identification of damage functions. Necessary, too, is extensive information about ecological processes in the evnironment. The processes between the various environmental media within one region and between different sub-areas, and the interactions between pollutants, again according to media and sub-areas, must be included in the system as well.

A comprehensive ecological accounting system would have to be drawn up separately for different regions because the composition and concentration of pollutants vary according to areas. Their interactions are also regionally different. Therefore, the damage they cause differs by region, too. However, the delimitation of regions for such ecological accounting systems is problematic. If we place the emphasis on the economic activities as causes of pollution, then the traditional criteria defining regional boundaries must be applied: an economically homogeneous area, or one determined administratively or historically. If, however, we look at the question in terms of environmental policy, then the emphasis must be on problem areas, such as boundaries of air basins and bodies of water in flow systems. The problem regions for the different environmental media overlap. Smog, for example, does not form according to county lines, so that administrative regions must be delimited by the extent of the problem areas.

The case for a regional orientation of such comprehensive ecological accounting systems should not detract from the necessity of including an examination of the importation and exportation of pollutants through the environmental media, such as air and water. Enough examples have been provided to indicate the danger of overlooking even minor interdependencies, which can lead to unforeseen overall effects.

Lastly, the period on which the system is based cannot be defined according to a calendar year because environmental problems vary significantly with the seasons. Therefore, depending on the regional conditions, it is recommended that seasonal accounting systems be devised.

Flow diagrams and pollutant balance sheets are methods of systematically describing the relevant processes in the environment. This description provides significant empirical information and can supply reference points for economic measures. It furnishes the economic decision makers with an idea of specific orders of magnitude.

IT ALL HANGS TOGETHER

"He jumps into the water and, all things being equal, he remains dry." This is, of course, an ironic illustration of the *ceteris paribus* clause. In more formal terms it concentrates on how one amount changes when a second varies—all other things remaining constant—and is an approach upon which scientists are often forced to rely. The *ceteris paribus* supposition cannot, however, be applied to the analysis of the relationships between the environment and the economy. The environment consitutes a complex structure of a multitude of interdependent subsystems like water, atmosphere, land, plants, fauna, cosmos, etc. Each one of the subsystems is in itself extraordinarily complex. For example, in a water system the various components, such as running water, still waters, ground waters, and seas are intricately interconnected. In addition, the subsystems interact with one another. The temperature of water affects air humidity, cloud formation and climate. Water evaporates and thus permits the fine particles in the water to enter the atmosphere. Rain and precipitation bring the pollutants from the air into the water systems. The differentiation between the various environmental media is a useful analytical tool, but it should not be permitted to disguise the fact that the various media are interdependent.

Unfortunately, too little is known about the extensive interrelationships between the environmental media. Hindsight is always 20/20. When Dr. Mueller discovered DDT and warned that its side effects had not yet been tested, no one could guess that a minute amount of DDT in water would multiply a million times within a short food chain because of its tendency to concentrate in fat. Why do we still not know whether the increase of sulfur dioxide in the atmosphere will eventually cool or heat the earth? How many other things can we not guess at or anticipate? The environment is extremely complex, and attempts to design representative models encounter enormous difficulties. To increase the labyrinth of interrelationships, humans must also be included somewhere, for this is the point of our study.

What are the effects of specific environmental qualities and specific pollutants on human health? The interdependence of environmental media means that pollutants cannot be assigned tidy pigeonholes in the various environmental media.

There are three reasons for this interdependence:

1) Substances in one medium can be transferred to another. Pollutants which are present in air, for example, can enter water through precipitation. Pollutants in water can enter the atmosphere through evaporation. These interrelating processes occur between all environmental media.

2) Methods of waste disposal are often substitutable. Many waste products can be disposed of in several environmental media, either discharged into the air or mixed with water, for example. If one is not permitted, the other will be exploited.

3) When one pollutant diminishes, another often increases. For example, the reduction of coarse dust in West Germany resulted in an increase in the much more hazardous fine dust, which serves as a bond for carcinogenic carbon monoxide and other dangerous particulates, such as toxins, and thus helps these substances to penetrate the lungs. A similar illustration concerns the automobile industry, where the installment of equipment in cars to limit the production of carbon monoxide has meant an increase in the still more lethal oxides of nitrogen. Substitutive processes of production, then, are also responsible for the interdependence of pollutants.

As a result of these interrelationships, economic policy measures must be applied to *all* environmental media. None can be left out. Every economic solution which does not take these aspects into account and which concentrates only on individual pollutants is irresponsible. A narrow policy may appease the public, but at the risk of increasing concentrations of pollutants whose effect could be worse.

The problem can be approached with the help of an input-output table. Originally developed by Leontief for economics,[15] it can be expanded to include the interdependencies between the economy and the environment. There are many interrelated sectors in the economy. Mining provides the iron and steel industry with coke, the steel industry furnishes automobile manufacturers with sheet metal, and the manufacturers produce cars for mining. An individual sector involves many goods from other sectors and delivers its products to consumers and various other market branches. An input-output table clarifies the mesh of relationships by listing the inputs and outputs of each sector. The upper left quadrant of Diagram 2 illustrates the output relationships between individual economic areas.

The environment can also be interpreted as a system of relationships in which one part, say phytoplankton, is the output of a process involv-

Diagram 2. Input-Output Table for the Economy and the Environment.

Inputs / Outputs			Economy						Environment				
			Sectors						Systems				
			1	2	3	.....	n	F[1]	1	2	3	....	m
Economy	Sectors	1											
		2											
		3											
		......											
		n											
		L[2]											
Environment	Systems	1											
		2											
		3											
		....											
		m											

F[1]: Final demand

L[2]: Labor

ing a combination of various factors, such as energy from the sun, nutrients, etc. On the other hand, phytoplankton is an input for other biological systems, such as fish. The input-output relationships in the environment can be seen in the lower right quadrant of Diagram 2. The columns list the inputs of a subsystem in the environment; the rows list the outputs.

In addition to the relationships existing *within* the economic system, and *within* the environmental sphere, there are also interactions *between* the two. The environment provides inputs for economic activities, as indicated in the lower left quadrant. Here might be listed the important elementary consumer goods of air and water, or cooling water for industry, water for irrigation, oxygen for combustion processes, and raw materials for production.

Lastly, the upper right quadrant presents the outputs from the economic system into the subsystems of the environment: warm water, which can no longer be used; waste products; pollutants. If the distribution of these outputs is known, such as the amount of phosphorus absorbed by zooplankton in a specific body of water, then so are the inputs.

Suppose sector 3 in the economic area presents the tanning and leather industries. Column 3, then, lists inputs into this activity, both from other sectors of the economy (skins, tannic acid, etc.) and those from the environment (water, oxygen). Row 3 presents the outputs of sector 3, again those delivered to the economic area (including final demand for such goods as treated leather) and those delivered to the environmental system (including used water, chemicals, and acids that are no longer usable).

The reader can well appreciate the extraordinary complexity of the breakdown illustrated in Diagram 2. In designing the table there are some relatively easy questions to be answered, like: How many sectors of the economy should be included? Or, how can we adequately distinguish between these sectors? But the problem becomes trickier when systematizing the environment. Which environmental media should be considered? How may innumerable sectors are there in the environment? Included are the multitudinous biological systems, the animal systems, etc. To impose a limit on otherwise numerically immense systems, they could be aggregated into groups. This poses the further question of whether aggregation is possible or useful, and whether it veils important information. The table must also acknowledge the substantial regional differences in input-output relationships between various economic and environmental areas.

The mesh of relationships between the economy and the environment illustrated in Diagram 2 can be interpreted in two ways. First, it can be an inventory for a past period, a systematic summary of observations on the environment and the economy drawn up at the end of a period, Such a comprehensive résumé could furnish valuable information. However, no such data exist. At best we can fill in only fragmentarily some columns or rows. Secondly, this mesh could be interpreted as a model of reality—not as a sum of observations, but as a respresentation of the interrelationships, as an explanation for the generation and production processes, or as a means of prognosis. It could be used to answer such questions as: How much zooplankton will be produced under what conditions? Or, how mych oxygen does this zooplankton produce? Or, where does this oxygen end up? Or, what effect does it have on other environmental processes?

There have been attempts to resolve the question of the interrelationships between economic branches through a relatively simple hypothesis: most production processes are so constructed that for the production of one unit of a good, a specific amount of various inputs is needed. For example, one ton of steel requires 0.0865 units from the primary production sector (energy and ore), 0.0256 units from the sector of steel, machinery, and automobile production, and 0.0180 units from wood, leather, paper, and textile production.[16] If production methods do not change significantly, these input-output coefficients can be used in answering such questions as how many inputs must the economy provide the steel industry in order to obtain a specific amount of steel? Unfortunately, such simple, linear relationships in the form of constant coefficients are found only rarely in the environmental system. Biological populations cannot be described by linear production functions, because the growth and adjustment processes involved are not linear.

Attempts at developing this input-output table not only indicate the great need for a systematic representation of all the interdependent factors, but they also show the sadly inadequate state of knowledge about the interdependencies between the environment and the economy. At least it outlines a direction for future research, and with the assistance of other methods such as simulation, some information on the relationships between the two spheres will become available.

The input-output analysis is of immediate use because it can provide information on the amount of wastes produced in certain processes and can help to identify which pollutants result from which economic activities. Suppose you buy a piece of furniture. We known more or less how much and which pollutants are generated by the furniture factory in its production process. Since the factory discharges almost none of the main air pollutants (carbon monoxide, sulfur dioxide, oxides of nitrogen, hydrocarbons, particulates), can we then conclude that furniture is an environmentally sound product?

No. The conclusion is false, because we must observe the furniture industry within the output mesh of the other sectors, examining how many inputs the industry uses from them. Thus, when we consider a piece of furniture we must imagine it within the input chain it constitutes, from the wood itself to the very slightly worn-down saw, from the wood stain to the depreciation of the equipment. If all the inputs into the piece of furniture are taken into account, our conclusion is bound to be very different. From the production of inputs required for one million dollars worth of furniture on the final market, the following pollutants can be identified: 46 tons of carbon monoxide, 31.1 tons of sulfer dioxide, 5 tons of oxides of nitrogen, 4.3 tons of hydrocarbons, and 11.7 tons of dust particulates.[17]

Leontief has calculated the pollutants involved in the final product of ninety sectors.[18]

Table 3 presents some of the results. The application of the input-output analysis represents an initial important source of information for economic policy measures. The model here only covers the five air pollutants, but it could analogously be designed to include other pollutants and other environmental media as well.

Leontief calculated the pollutant content of one unit of final demand. Other authors[20] have reported the pollutant content related to one unit

Table 3.[19] Pollutant Content of Final Products

	Pollutants				
Sectors	Carbon Monoxide	Sulfur Dioxide	Nitrogen Oxides	Hydro-carbons	Particulates
Electricity Plants	11.4	788.8	185.0	3.5	295.5
Paper Mills	32.0	40.0	32.0	9.7	177.0
Refineries	108.0	127.0	5.7	46.7	13.9
Non-Iron Metals (Primary Stage)	23.0	1038.0	13.5	4.8	58.6
Limestone Industry	14.5	43.5	10.5	4.3	2191.7
Household Goods	46.0	31.1	4.9	3.6	11.7
Air Transport	312.0	21.7	2.8	43.0	4.9
Water Transport	103.7	103.0	60.8	32.7	35.0

of net value added. This information is often used in decisions about industrial location and promotion, as in development policies for underdeveloped areas. The net value added, i.e., the income generated by the factors of production, is an indicator of the desirability of an area for settlement or development. The sectors of the economy accorded most significance in development planning are those which (in addition to other characteristics, such as effects on employment and balance of payments), promise a high net value added for the factors of production employed. If the environment is included in calculations, then it will be necessary to account for the pollutants generated by each sector, too. By relating the units of pollution to the net value added by each sector, economic policy makers are forced to choose between the conflicting objectives of economic growth and improving environmental quality.

NOTES

1. Council onEnvironmental Quality, Annual Report 1974, pp. 267–276. The annual reports of the CEQ are an interesting source of statistical data on pollution. The reader is also referred to the survey mentioned in Chapter 2 by Allen V. Kneese, Economics and the Environment, Middlesex, 1977.

2. Battlelle-Institut, Frankfurt am Main, Schaetzung der monetaeren Aufwendungen fuer Umweltschutzmassnahmen bis zum Jahre 1980, Umweltbundesamt, Bericht 1/76, September 1975, Berlin, p. 31.

3. Source: Council on Environmental Quality, *Annual Report,* 1973, p. 6.

4. Council on Environmental Quality, 1973, p. 266.

5. Bundesminister des Innern, Materialien zum Umweltprogramm der Bundesregierung, 1971, p. 207.

6. Council on Environmental Quality, 1972, p. 11.

7. "Orama der Ozeane," *Bild der Wissenschalt* 7, 1977, p. 54.

8. A. V. Kneese, Economics and the Environment, Middlesex, 1977, p. 44.

9. Bundesminister des Innern, op. cit., p. 129.

10. Ibid., p. 132.

11. G. Olschowy (ed.), *Belastete Landschaft—Gefaehrdete Umwelt,* Munich, 1971, p. 23.

12. See Chapter 8.

13. "Mehr Freiheit, mehr Konservendosen—Spiegel-Report ueber die Muelllawienen in Bundesrepublik," *Der Spiegel* 49:73(1971).

14. A. V. Kneese, "Environmental Pollution: Economics and Policy," *American Economic Review,* Papers and Proceedings 61:155–177 (1972).

15. W. Leontief, *Input-Output Economics,* New York, 1966; with reference to the application of the approach to ecological problems see W. Isard et al., *Ecologic-Economic Analysis for Regional Development,* New York, 1972.

16. R. Krengel et al., *Input-Output Relationship for the Federal Republic of Germany,* 1954–1960, Berlin, 1969, Table C-1.

17. W. Leontief and D. Ford, "Air Pollution and the Economic Structure: Empirical Results for Input-Output Computations," in A. Brodey and A. P. Carter (eds), *Input-Output Techniques,* London, 1972.

18. Ibid.

19. Source: Ibid., pp. 9–30.

20. J. C. Hite et al., *The Economics of Environmental Quality,* Washington, D.C., 1972.

Chapter III

Itai-Itai Smog and Killer Fog: The Environmental Damages

Sooner slays ill air than sword

Anon. Ratis Raving, c. 1450

People of the Jingtsugawa Valley in the Japanese prefecture of Toyama suffer and die from a chronic cadmium poisoning called the itai-itai sickness which results from high concentrations of cadmium in the body. This toxin, which can be traced to industries producing or using cadmium such as mines, foundries and battery factories, causes the kidneys to malfunction, hinders the building of calcium, an important element in bone development, and destroys the calcium balance of the body. The resulting osteomalacia can be fatal.[1]

According to the AMA, an increase in deaths and illness has been documented in periods of high air pollution. The consequences are particularly severe for the aged and for persons with respiratory ailments such as asthma and bronchitis, and those with heart and circulatory problems.[2]

Dr. Wilbert Aronow studied the effects of air pollution on heart patients. He drove ten angina pectoris patients through Los Angeles traffic for one and a half hours. Alternately he allowed them to breathe the polluted city air and fresh air from containers. When breathing the city air the patients registered lower blood pressure, slower pulse, a reduction in the oxygen content of the blood, and an increase in the proportion of poisonous carbon monoxide in the blood.[3]

The deaths caused by the killer smogs in Belgium, Donora, London, and Poca Rica, and the Seveso tragedy in Italy contribute further testimony to the adverse effect that acute air pollution episodes have on human life. There are also many indications that long-term exposure to increasing environmental pollution takes its toll on human life.

Only by considering the background of environmental damage can the question of environmental pollution be discussed adequately. It is pointless to talk of environmental policy measures without sufficient knowledge of the order of magnitude and the types of damages incurred. An examination of the extent and kinds of damages involved ought to indicate whether the interest in environmental problems is only a passing fad, or whether it is a discussion to which increasing attention will be paid in the future.

We will distinguish between five different types of damage: health damages, ecological damages, damages to the environment as a consumer good, property damages, and production damages. This theoretical classification must not disguise the actual interrelations between the various types of damages. Damages to health, for example, and effects on the environment as a consumer good are obviously related to ecological degradation.

FRAGILE: HANDLE WITH CARE

Many scientific studies have been conducted on the effects of pollutants on human health and mortality. To present all the results of these analyses would require an entire book. This discussion, therefore, is restricted to reporting some significant findings and to mentioning a few sample cases.

Environmental damages can be analyzed according to the effects on one individual or statistically for many people. Both cases deal with persons affected by pollution, either by a very high concentration for a relatively short period of time, as exemplified by the killer smogs during an atmospheric inversion, or by a lower concentration of a pollutant over an extended period. Medical examinations based on an individual or a small group usually have the advantage of being able to obtain a detailed diagnosis, and thus are fairly sure of establishing a causal relationship between an ailment and a specific pollutant. Statistical studies on many cases, on the other hand, benefit from a broader statistical base, but sickness or death diagnoses become less certain. Some results from the multitude of studies conducted are presented below.

1) Lave and Seskin[4] conducted a study of 117 cities in the United

States and found a statistical relationship between death rates and air-pollution levels. Basing their study on long-term concentrations of pollutants, rather than acute periods, the authors concluded that a 50 percent reduction in air pollution would raise the life expectancy of humans three to five years. What this means for the individual reader and his or her children is painfully obvious.

2) The two authors arrived at the same conclusion in another statistical study[5] on the relationship between high, short-term daily concentrations of pollutants and daily deaths occurring in Chicago, Denver, Philadelphia, St. Louis, and Washington, D.C. A direct correlation was found between the effects of pollution and the deaths recorded in Chicago during the years 1962–1966. Lave and Seskin explain that such a correlation could not be established in the other cities because the short-term concentrations there were significantly lower than in Chicago. For example, pollutant concentrations were ten times higher in Chicago than in Denver.

3) The incidence of lung cancer in men living in the West German cities of Dortmund, Duisburg, Duesseldorf, Essen, Gelsenkirchen, Muelheim, and Wuppertal has risen sharply since 1960. Five times as many deaths attributable to lung cancer are registered in the densely populated areas of the Ruhr industrial valley than in the countryside.[6] In the United States it has been reported that the death rate both among smokers and nonsmokers is higher in cities than in rural areas.[7] In the last ten years New York has recorded a fivefold increase in the number of deaths caused by emphysema, a swelling in the lungs which hinders the functioning of the tissues needed for breathing.[8]

4) A study conducted in New York during 1962–1964 indicated that more deaths occurred in periods of heavy air pollution, specifically when high levels of sulfur dioxide and particulates were recorded.[9]

5) During a seven-day period characterized by high air-pollution levels in November 1966, 168 more people died in New York than in comparable periods. The number of bronchitis and asthma patients also increased.[10]

6) An expert commission revealed that 2 percent of the population in a nonpolluted area reported symptoms of chronic bronchitis, 5 percent in a moderately polluted area, and 10–15 percent in a heavily polluted area in the United States.[11]

So much for statistics. A German authority on industrial medicine states, "the acute, short-term effects of very high pollutant concentrations during an atmospheric inversion have been proved. In addition to higher death rates, other effects on health have been recorded, for example in cases of bronchitis and heart ailments."[12]

As unequivocal as this evidence may seem, however, there are complications in its interpretation. The second study by Lave and Seskin cited above is an example of the problems which arise. The study was based on the atmospheric concentrations of carbon monoxide, sulfur dioxide, oxides of nitrogen (NO, NO_2), and hydrocarbons. The researchers devoted special attention to fluctuations in the death rate. For instance, did unusually high pollutant concentrations cause an increase in deaths over a few days, and did the death rate then sink below the average after such days? This did not prove to be the case. Short periods of intense air pollution did raise the death rate, but the rate did not subsequently fall below the average.

The problem with this regression analysis lies in the unequivocal isolation of air pollution as the causal factor of deaths. The mortality rate is not influenced by air pollution alone. The age and sex distribution of the population, genetic factors, nutrition, smoking habits, sports, income, ethnic group, quality and frequency of medical consultations, employment, and climatic factors also play a role. If these factors could be held constant, allowing only the air pollution levels to vary, it might be possible to determine the relationship between death rates and air pollution. Since these empirical conditions can almost never be found, attempts are made to "calculate out" the other influencing factors. Such studies are questionable because many of the above-mentioned determinants (such as genetic factors) cannot be specified, the classification of certain sicknesses and causes of death may be false, and a person might not die in the place he or she has been exposed to extended periods of air pollution (if the person moves or changes jobs, for example). In spite of the doubts of theoreticians, however, these results provide food for thought until proven false.

The results of individual experiments based on specific pollutants, particular ailments, and often only a few patients, are also shocking.

7) The minimata disease is a mercury poisoning which causes lethal paralysis, leads to bronchitis, and destroys air sacs in the lungs. The symptoms are open wounds in the mouth and gums, inflammation of the gums, loss of teeth, paranoia, and personality changes.[13]

8) The effects of radioactive rays have also been observed in experi-

ments with various chemicals. Certain chemicals contained in herbicides and pesticides belong to this group of radiomimetic elements which can damage chromosomes, hinder normal cell division, or cause mutations. Experiments with animals have proven many such chemicals to be teratogenic, i.e., they cause deformation in the young.[14]

9) Little is known about the effects of DDT on humans. High concentrations of DDT can incur watery eyes, severe perspiration, impaired vision, and diarrhea. Exceptionally heavy dosages cause respiratory problems and purple discoloring of the skin. Since DDT accumulates in fatty tissues of the body, it is necessary to research the long-term effects, concentrating, for instance, on the effects produced when these tissues are dismantled during illness, allowing the accumulated DDT to enter the blood.[15] Studies of wild animals have shown effects extending to the second and third generations.[16]

10) Polycyclical hydrocarbons such as 3.4 benzyprene—like so many pollutants, the results of incomplete combustion—enter the lungs with fine dust. .02 mg, a mere thousandth of the 20 mg which collect in the human body in sixty years, suffices to induce tumerous growths when injected under the skin of animals. Although benzyprene is eventually decomposed in the lungs, it must be remembered that all cancer-inducing irritants inflict injury which cannot be healed and which worsens with each new irritation.[17]

11) Carbon monoxide affects the central nervous system and the cardiovascular system. The effects range from slight disturbances in time perception, impaired alertness, at a CO content of 2–2.5 percent in the red blood pigment, to problems with concentration, severe drowsiness, and decreasing ability to distinguish light at a CO content of up to 4 percent. Physical disability and finally poisoning occur in cases of CO levels exceeding 9 percent. Further, at CO contents between 5 and 9 percent, a weakening of blood circulation through heart muscles, a deterioration of the myocord, and irregular pulse in heart patients have been observed. A significant statistical relationship has been found between fatal heart attacks and high CO concentrations in the atmosphere.[18]

12) Health is threatened when the average annual concentration of sulfur dioxide in the atmosphere is higher than 0.04 ppm. Deaths resulting from bronchitis and lung cancer increase when this level of sulfur dioxide combines with dust concentrations of around 0.06 ppm. These

levels are often exceeded in American cities. Negative effects on health have been observed when concentrations of sulfur dioxide exceed 0.11 ppm three or four consecutive days, a situation which often develops during atmospheric inversions.[19]

13) Multivariant regression analysis indicates statistically significant relations between cancer mortality rates in Louisiana and drinking water obtained from the Mississippi River. This is particularly true of cancer of the urinary organs and cancer of the gastrointestinal tract.[20]

The great epidemics which in previous times raged through cities like brushfires have largely been brought under control today. But these studies indicate a contemporary variation that is more insidious—the chronic illness. Unlike the infectious diseases of the past, modern pollutants spread secretly, often unnoticed, slowly accumulating in our bodies. The result is a chronic ailment, the continuing malfunctioning of organs which often leads to an early death. Heart and circulatory disorders were responsible for half the deaths in the United States in 1962. Lung cancer, once a rare case, now kills more people than do all other types of cancer together. Emphysema patients in the United States doubled in number every five years since World War II. A causal relationship has been established between air pollution and these diseases as well as others, such as asthma, bronchitis, respiratory tract ailments, allergies, and many childhood illnesses which may be the causes of later chronic problems.[21]

NATURE'S CYCLES

Closely connected to the effect of pollutants on human health and life is the damage they cause to the natural environment. Thanks to recent research by ecologists, we are gradually coming to recognize that the environment is a highly complex mechanism composed of an incredible multitude of interrelated cycles.

A key characteristic of these cycles is homeostasis—the ability of the individual subsystems of nature to maintain a balance and to return to the equilibrium after minor disturbances. The fish-bacteria-algae cycle is a good illustration. The fish release organic wastes which are converted to inorganic products by the bacteria. This inorganic matter is food for the algae, which in turn are eaten by the fish. This cybernetic system can regulate itself within certain limits. For example, a short-term increase in algae is kept under control by an increase in the fish population.

However, if the cycle is seriously interfered with, it can collapse. When too great an amount of phosphates and nitrates is dumped into the water, it causes a greater increase in algae than the fish can control. A runaway growth in the algae means a like increase in the dead algae that must be decomposed by bacteria. Since these bacteria need oxygen, the oxygen level of the water falls, and the fish die. In extreme cases, the body of water dies. Humans have certainly interfered with this system of cycles in many ways. In addition to dumping solid wastes, they have burdened the environment with totally new substances like plastics, toxic metals, herbicides, pesticides, and dozens of chemicals.

The complexity of environmental systems is not only due to the delicate balance of the individual cycles, but stems also from the fact that many of these cycles are interrelated through common variables. We researchers are too accustomed to observing isolated phenomena in our scientific experiments. We research the effects of a new pesticide, such as DDT, on the malaria mosquito, but do not analyze the potential side effects of such a pesticide, do not study the process by which minor side effects snowball into serious, unforeseen cumulative results. We observe individual phenomena and hope that the sum of our observations will yield the whole picture.

Ecological damages are not determined so much by the disappearance of individual animal species, nor by the gradual extinction of amphibians in Lake Erie. The ecological damage lies in the relationship of these individual environmental components to the cycles of nature, to homeostasis, to the equilibriums within each system. And each one of these equilibriums defines the human environment. Damages caused to ecological systems therefore also affect people. In this way, the objective of environmental conservation becomes instrumental in achieving the principal societal goal of raising the standard of living.

The conservation of the environment is therefore an objective of economic and social policy in its own right, and it is high time that it be accorded a place beside the other goals of full employment, price stability, balance of payments, and economic growth. Economy and ecology have much more in common than their first three letters. Environmental conservation must be included in the calculations of economists. The problem with assessing ecological damages lies first in the fact that the relationship of the individual environmental components to the whole system is very difficult to establish. Secondly, many ecological damages are long-term, appearing only after ten or twenty years. Such damage is hard to identify and not easily attributable to a specific pollutant. It is often argued that since ecological damages cannot be foreseen, they cannot be taken into consideration in present planning procedures. This reasoning is unacceptable, for even when no information at all is available about possible damages, we must at least include the risk of such damages in the economic calculations. If the risk factor were included as a social cost in a decision regarding, for example, the manufacture of a certain chemical, then it is quite possible that such a product would no longer appear profitable.

Admittedly, there is a great deal of information yet to be obtained on the subject, and we must deal with vague ideas about potential damages. But a significant amount of knowledge is already available about certain types of damages, providing a useful basis for decision making. Thanks to the work of ecologists we already have many hypotheses which at least permit conjectures about risks. For example:

1) Carbon dioxide in the earth's atmosphere plays a central role in the ecological system. Like glass, this gas allows the sun's rays to pass to the ground, where they are absorbed as energy, but it reflects the infrared rays returned by the earth. In these greenhouse-like conditions, the temperature of the atmosphere rises. The normal carbon dioxide content of the earth's atmosphere is 0.03 percent. The level has increased by 20 percent since the turn of the century. If we continue our extravagant consumption of fossil fuels, the carbon dioxide level will double within forty years.[22] It is easy to imagine how a change in the average temperature would affect the earth's equilibrium, especially with regard to the polar ice caps and, consequently, sea levels.

On the other hand, the greenhouse effect is welcomed by those scientists who point to the alternative hypothesis which holds that the increase in particles and aerosols permits less energy from the sun to reach the earth, therefore causing the earth to cool down.

The inability of experts to disprove one or the other of these contradictory hypotheses is doubtless disturbing to the layperson. Both alternatives reckon with negative effects, albeit different ones. Since we cannot be certain that the two types of effects will cancel each other out exactly, the risk to humans of a change in the earth's atmosphere must be included in economic calculations.

2) Oxygen was lacking in the primeval atmosphere of the earth. It was produced after a three-billion-year process through the photosynthesis of slowly developing plants. We are now threatening the earth's store of oxygen. The demand for oxygen is constantly rising. Every combustion process requires oxygen. A VW uses as much oxygen in 900 km as does a

person in a year; a jet at takeoff consumes as much as a 17,000-hectare forest produces in a night.[23] Combustion processes in Switzerland use up as much oxygen in a year as is needed by 81 million people.[24] The rising demand is confronted with a decreasing supply. The bodies of water, especially the oceans, our major oxygen producers, are being polluted. The American geologist Wuster believes that a concentration of 0.1 ppm of DDT in the ocean is sufficient to reduce photosynthesis by phytoplankton. And the enormous forests, such as those in the Amazon area, Canada, and Siberia are being felled more intensively.[25]

3) The ozone layer which encircles the earth at an altitude of between 9 and 35 miles protects life from ultraviolet rays.[26] This layer is being destroyed by propellants, especially fluorohydrocarbons from spray cans. This is yet another example of a new product which is highly effective for its intended purpose, but whose side effects cannot be foreseen.[27] These chemicals were introduced in the 1920s as a cooling agent. They are well suited to use in spray cans since they are chemically inert and therefore do not interact with the substance to be sprayed. This characteristic, however, permits the chemical to rise higher and higher in the atmosphere without being impeded by other substances. Ultraviolet rays cause chlorine atoms to be released, transforming the ozone into simple oxygen. Nitrogen oxides, waste products from high altitude aircraft, cause similar damage, although to a lesser extent. Experts suspect that this dilution of the ozone layer can have such long-term consequences as skin cancer, biological permutations, or even death.

4) The California sardine industry, with its legendary catches of 800,000 tons, is gone. There are no sardines left. At first, the small catches were attributed to overfishing in previous years—until a biologist noticed that the fish population was not reacting as expected. A severe drop in a natural community's population is usually followed by an immense increase, until the equilibrium is restored. This did not occur with the sardines. How could this unexpected reaction in nature be explained? A study of the amounts of DDT employed in California showed that the sardine schools decreased in proportion to the increase in DDT used in the state.[28]

5) Overfishing is another serious form of human interference in the ecological equilibrium. In the Humboldt stream off the coast of Peru, for example, 30,000 tons of anchovies were caught in 1953. The catch rose to 12 million tons by 1970. The number of people engaged in fishing increased from 2,500 in 1957 to 20,000 in 1966. The experts of the Food

and Agriculture Organization of the United Nations warned that these enormous catches would destroy the capacity of fish to regenerate themselves. But no one listened. In 1972 the fish began to disappear. The catch fell from 10 million tons per year to 4.4 million, which represented a reduction of 10 percent of the world catch. In 1973 the total catch was only 2 million tons. Changes in the course and the temperature of the Humboldt stream aggravated the effects of the uncontrolled fishing, and the whole industry collapsed. Since these fish were not only a part of an economic system, but also of an ecological cycle, the effects of this loss were felt in other ways, too. The millions of birds which had depended on the anchovies for food disappeared. These birds had been leaving very rich excrement, called guano, on nearby islands. This product had been sold by Peru as high-grade fertilizer and had been a major source of foreign exchange. Twenty thousand fishers and 8,000 fish-meal factory workers lost their jobs. In 1974 the government of Peru introduced fishing restrictions, and improvements have already been registered. The production has increased again to approximately 4 million tons.[29]

6) It can be expected that the extraction of seabed resources, such as manganese nodules, will affect the ecological balance of the oceans. The extraction process itself may be pollution intensive, and the processing of the minerals on board the ships (e.g., bleaching) may cause new emissions and a reduction in the quality of the marine environment.

7) Starfish are disturbing the equilibrium of many ocean communities and are gnawing away entire coral reefs which protect Pacific islands from the pounding waves. Presumably, the natural predators of the starfish eggs have been destroyed by DDT, permitting the starfish to multiply unhindered.[30]

8) The Caspian Sea is so polluted that the white salmon have died off and sturgeon have been reduced by 60 percent.[31]

9) A millionth of a gram of fluoride in a cubic meter of air can cause damage to plants, whose leaves absorb the substance. Cattle which assimilated the fluoride contained in the plants suffered fluoride poisoning that resulted in tooth and bone changes and severe weight loss.

10) The sulfur dioxide content usually measured in our city air causes plants to undergo a chronic damage referred to as early aging. Oxides of nitrogen have a similar effect, and seem to impair the growth of plants even when no external damage can be seen.[32]

11) A full two thirds of the ponderosa pine in the San Bernardino Mountains of Southern California have either died or withered away.

Experiments have shown that this phenomenon can be attributed to the ozone content of Los Angeles smog, which has now spread far enough to reach these forests at an altitiude of 1,000 meters. If the forests die, not only will the beautiful landscape vanish, but the soil and moisture held by the roots of the trees will be lost.[33]

12) Red tides are a relatively new phenomenon in our oceans. They are caused by a sudden blossoming of certain types of phytoplankton, such as *Noctiluca scintillans,* and are often accompanied by significant numbers of dead fish. They have been observed after violent storms, such as typhoons, which churn up from the depths the nutrient-rich chemicals which then swirl up to the upper layers of water. Simultaneously, a reduction in the salinity of the coastal waters through the heavy rainfall seems to create favorable conditions for the growth of these types of plankton. Biologists are interested in the effect this red tide has on fish populations and on the ecological equilibrium of the seas.[34]

Reams could be written if all the ecological damages were to be listed.[35] We have limited ourselves to a few examples. It should be clear that these damages have direct consequences for the human environment. Only too often are they accompanied by health problems.

POLLUTION IN DOLLARS AND CENTS

Closely connected to health and ecological damages are the damages to the environment as a public good. The environment is used for various types of consumption; air and water for primary consumption, nature for physical and spiritual regeneration through swimming, hiking, mountain climbing. The use of the environment as a recreational medium ranges from immediate recreation, such as relaxation after work in the neighborhood, to intermediate recreation on weekends in the surrounding countryside, to extended recreation during vacations.

Environmental services are always defined by many characteristics. Immediate recreation, for example, is defined by the quality of the air, traffic noise, availability of parks—all of which are, in turn, composed of different components. Pollutants change the quality of the environment. In degraded areas they make breathing difficult and render swimming unthinkable. In short, they affect the value of the environmental services and often make consumption of these services impossible.

Water pollution changes the taste of drinking water and creates foul-smelling lakes which rob a walk of its romantic charm. Smog, smoke, and haze reduce visiblity. In Los Angeles, where fifty years ago you could see forever if the weather was clear, summer visibility is now only three miles, if that. The refreshing beauties of nature are being veiled.

Other aesthetic values must be considered in this context. The landscape changes through construction and overconstruction. The glorious freeway system, the constant drone of traffic, only add to the list of factors which affect the use of environmental services.

Environmental pollution can also affect production processes by degrading raw materials such as forests and fish populations. Property damage represents the fifth category of environmental damages. It can appear both in productive capital and in durable consumer goods. Property damages are indicated, for example, by repair and upkeep costs for buildings and equipment, by drops in property value or rent. Since fewer people want to live in a highly polluted area, a freely functioning housing market in which there is no excess in demand will reflect a drop in the value of real estate.

Here are some examples of production and property damages and damages to the environment as a consumer good.

—European monuments are being destroyed by the corrosive air we breathe. The Colosseum in Rome had to be closed temporarily because it had suffered severe damages from air pollution. The façades of the Cologne Cathedral are decaying faster than the stonemasons can repair them.[36]

—The fluoride and sulfur dioxide emissions from the phosphate mines in Polk County, Florida, have harmed both citrus groves and animal husbandry. This has clearly been reflected in the marked fall in the price of agricultural land, which is determined by the expected level of productivity.[37]

—Sulfur oxides attack the hardest and most durable surfaces. Steel rusts two to three times faster in industrial centers than in rural areas. When particles are present in the air the rate of corrosion is magnified. A third of the wear on railway tracks in Great Britain is attributed to sulfur oxides.[38]

—Particulates cause significant problems of upkeep, such as cleaning façades, frequent painting, washing of cars, clothes, etc. Ozone damages textiles, bleaches colors, and accelerates the deterioration of rubber.[39]

—Over a fifth of the mussel and oyster beds in the United States had to be closed on account of water pollution. Before 1935 between 1 million and 3 million pounds of crab were caught annually in San Francisco alone. This industry no longer exists. The annual prawn catch dropped from 6.3 million pounds before 1936 to 10,000 pounds in 1965.[40]

It is incredibly difficult to express the compromise of aesthetic value in dollars and cents. How much has been lost when you can no longer fish in Lake Erie? When trees die? When works of art are destroyed? When human health is compromised? When people die? The following estimates have been made of damage caused by air pollution in the United States:

—The costs for cleaning façades of buildings in two cities with different levels of air pollution were compared. It was found that in the city with greater air pollution $84 per capita more were spent annually for this purpose than in the other city.[41]

—Because of air pollution damage, $100 million are spent each year on painting steel construction.

—Commercial cleaning, washing, and dyeing of textiles damaged by air pollution are estimated to cost $800 million.

—The damage incurred to agriculture, plants, and domestic cattle totals over $500 million.[42]

—According to the Council on Environmental Quality, corrosion and other damage caused to trees, plants, and materials by air pollution are estimated at $4.9 billion each year.

—A study on property damages concluded that air pollution has reduced the value of buildings and land by $5.2 billion.

—If we try to account for health damages attributable to air pollution based on hospital and doctor's bills, and if we add (how macabre!) the loss of income due to illness and death caused by air pollution, then yet another $6 billion are added to the costs of air pollution.

The total costs, then, of damage to health, vegetation, materials and property add up to $16 billion annually, which is over $80 per person.[43] And that is only air pollution!

These are estimates provided by a government agency which admits the figures are *under*estimated. They take into account only the hospital costs and lost income, but not the costs borne by the patient and family in terms of discomfort, fear, uncertainty, despair, depression, and frustration. Neither aesthetic values nor the ecological damages are represented in the above figure. And, presumably, the estimates exclude some damages to health, production, and property because we still know so terribly little about the effects of certain pollutants.

THE DAMAGE FUNCTION

The observation and identification of damages, which in the past were related to a specific amount of a pollutant, have now become significant information for economic policy making. Environmental policies, however, require totally different information. For example, how does the damage vary when the amount of a pollutant increases or decreases? Looking at the problem graphically, this means that in a diagram with designated axes for damages and amounts of pollutants, we must know not only point A, but also the damage function. Diagram 3 illustrates such a damage function.

Damage functions are a relationship between individual pollutants such as carbon monoxide, particulates, etc., on one side, and the damages on the other. The damages can thereby be expressed in a wide variety of values such as the number of bronchitis cases, the stages of illness, the stages of eutrophication, the probability of atmospheric changes, the time needed to drive to a suburban home in order to escape the city air, or finally, in monetary terms, like decreased production and real estate prices.

Damage functions are largely unknown. At best we know something about individual points on the graph. The knowledge of damage functions provides extremely important information for the resolution of pollution problems. They are a focal point in environmental policies. It is to be expected that individual pollutants will have very different effects. We know, for example, that fluoride has the same effect on plants as a hundred times greater concentration of sulfur dioxide. The damage function for fluoride is thus very different from that for sulfur dioxide. Therefore, economic measures to deal with a ton of fluoride also have to be significantly more stringent than those designed to deal with a ton of sulfur dioxide.

Determining damage functions entails many problems, some of which will be discussed here.

1) The damage functions of very diverse pollutants must be ascertained.

2) The damage function of a particular pollutant will differ signifi-

cantly according to the environmental medium involved. One reason for this is that the different media are utilized for different purposes. Another reason is that the pollutant can react differently per amount employed in the different media.

3) It can be assumed that damages increase with a greater amount of pollutant. In Diagram 3, four possible curves are illustrated. Curve *a* represents the case in which increasing amounts of a pollutant cause increasing damage, with the latter gradually leveling off. Curve *b* illustrates a linear rise of damage with added amounts of a pollutant. Curve *c* represents the case of progressively increasing damages. Curve *d* assumes that after a certain threshold has been crossed, the damage a pollutant causes suddenly jumps. Which of these courses describes reality remains a totally unresolved question.

4) The course of the curves in Diagram 3 depends on when the assimilative capacity of the environment is exceeded. Let us suppose that a portion of the pollutant is decomposed by the environment. If the assimilative capacity of the environment is exceeded, a severe increase in the damage function (discontinuity) is to be expected. If a body of water dies, for example, then those pollutants which had formerly been assimilated can no longer be decomposed. In such cases, then, curve *d* is a plausible illustration of the damage function.

Diagram 3. Damage functions

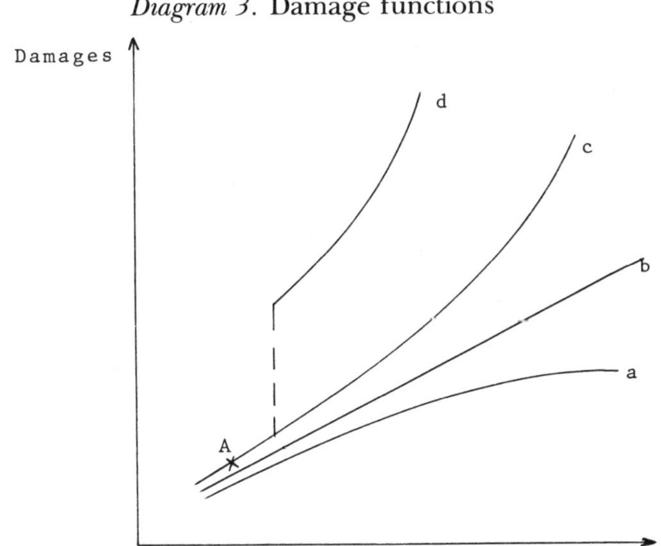

5) Some pollutants can be decomposed, such as organic matter in water; others, however, are persistent. The course of the damage function also varies according to the degradability of substances. In the case of degradable materials, basically only the pollutants that are introduced into the medium during a given period are of interest to policy makers, and they can simply be plotted on the horizontal axis. On the other hand, nondegradable pollutants present a very different picture. These pollutants accumulate over time, so that during a given period not only do the newly discharged pollutants have to be considered, but so do those pollutants still remaining from previous periods. As a rule, these are chemical substances, because, in contrast to nature, chemistry does not provide a process to decompose each new substance it produces.

6) Pollutants can interact with each other and thereby produce a new pollutant which may cause even greater damage than the original pollutants alone. One example of such synergism is the Los Angeles smog which develops from the interaction of the sun, oxides of nitrogen, and unburned carbon monoxide. Another example is the generation of sulfuric acid from water particles and sulfur trioxide. These interactions would also be represented on the graph as a discontinuity.

7) The relationship between damages and amounts of pollutant is not identical to a relationship between damages and discharged wastes. Waste is not a pollutant until it cannot be assimilated by the environment. On account of natural diffusion processes, interactions between pollutants, and the assimilative capacity of the environment, the pollutants present in an environmental medium do not equal the wastes emitted into the environment. The nature and extent of the environmental damages are determined by the amounts of pollutants. In order to reduce overall pollution, policy measures must be directed at the wastes discharged into the environment.

This distinction has important consequences for the formulation of economic policies. It is absurd to tax individual enterprises on the amount of waste they discharge without first determining the amount of pollutants present in the environmental media. A medium in a certain region may already be so overloaded with accumulated pollutants that the addition of even small amounts could cause a discontinuity in the damage function.

8) The total damage increases with the number of people affected. This obvious statement throws light on an important problem. As a result of population growth, the spatial concentration of the population in

urban centers, and the high industrial development near residential areas, a given amount of pollution must be borne by more and more people simultaneously, thus shifting the damage function upwards.

NOTES

1. A. V. Kneese et al. (eds), *Managing the Environment, International Cooperation for Pollution Control,* New York, 1971, p. 189.
2. *Journal of the American Medical Association:* 190 (1972).
3. Los Angeles *Times,* Dec. 3, 1972; *Der Spiegel* 52: p. 103 (1972).
4. L. B. Lave and E. P. Seskin, *Air Pollution and Human Health,* Baltimore, Johns Hopkins University Press, 1975.
5. L. B. Lave and E. P. Seskin, "Acute Relationships among Daily Mortality, Air Pollution, and Climate," Lecture at the Conference on the Economics of the Environment, organized by Resources for the Future and the National Bureau of Economic Research, Chicago, Nov. 10 11, 1972.
6. "Siedlungsverband Ruhkohlenbezirk, Ein Planet verrottet," paper on environmental protection, Essen, 1971, p. 112.
7. A. A. Haesler, *Mensch ohne Umwelt,* Olten and Freiburg, 1972, p. 94.
8. Ibid.
9. J. McCarrol and W. Bradley, "Excess Mortality as an Indicator of Health Effects of Air Pollution," *American Journal of Public Health,* 1966.
10. M. Glaser et al., "Mortality and Morbidity During a Period of High Levels of Air Pollution," *Archives of Environmental Health* 16, 1967.
11. OECD, Report of the Air Management Sector Group, 1974, p. 23.
12. H. W. Schlipkoeter, "Die Luftverunreinigung als gesundheitliches Problem, in H. Schultz (ed), *Umweltreport,* Frankfurt, 1972, p. 200.
13. "Mehr Freiheit, mehr Konservendosen, Spiegel-Report ueber die Muellawine in der Bundesrepublik," *Der Spiegel* 49:62 (1971).
14. F. Baehr and W. Gruner, "Hygienisch-toxicologische Aspekte der Umweltsituation." in H. Schultz (ed.), *Unweltreport,* Frankfurt, 1972, p. 287.
15. D. Widener, *Kein Platz für Menschen,* Franfurt, 1972, p. 31.
16. J. Borneff, "Forderung and die Reinhaltung unserer Binnengewaesser aus hygienischer Sicht," in H. Schultz (ed.), op. cit, p. 88.
17. H. W. Schlipkoeter, *Umweltreport,* op. cit., p. 203.
18. Ibid., p. 201.
19. Council on Environmental Quality, 1971, p. 68.
20. T. Page, R. H. Hamis, and J. S. Epstein, "Drinking Water and Cancer Mortality in Louisiana," *Science* 193:55–57.
21. Council on Environmental Quality, 1970, p. 68.
22. A. A. Haesler, *Mensch Ohne Umwelt,* op. cit., p. 105.
23. H. Strohm, *Umweltschutz,* Darmstadt 1972, p. 72.

24. A. A. Haesler, p. 87.

25. C. Wuster, "DDT Reduces Photosynthesis by Marine Phytoplankton." *Science:* 159.

26. *Time,* Feb. 23, 1976.

27. The human imagination cannot anticipate all possible effects of a new substance. This is true of ecological effects as well as behavioral changes of humans and institutions. The equipment installed in American bars to measure the alcohol level of the blood was not intended to encourage drinking competitions, as has been observed in many cases.

28. D. Widener, *Kein Platz für Menschen,* op. cit., p. 38.

29. "Das Drama der Ozeane" *Bild der Wissenschaft,* Nr. 7, 1977, p. 49.

30. D. Widener, *Kein Platz,* op. cit. p. 44.

31. A. A. Haesler, *Mensch Ohne Umwelt,* op. cit., p. 78

32. Council on Environmental Quality, 1970, p. 70.

33. D. Widener, op. cit., p. 76.

34. B. Morton and P. R. Twentyman, "The Occurrence and Toxicity of Red Tide Caused by *Noctiluca scintillans* . . . in the Costal Waters of Hong Kong," *Environmental Research* 4, 1972, pp. 544 et seq.

35. For more information on the ecological damages, see, for example, B. Commoner, *The Closing Circle,* New York, 1971, p. 29

36. U. Zuenddorf, *Untergang in Raten,* Düsseldorf, 1972, p. 41.

37. Council on Environmental Quality, op. cit., 1970, p. 70; T. D. Crocker, Externalities, Property Rights and Transactions Costs: An Empirical Study, Journal of Law and Economics, Vol. 14 (1971) p. 451–464.

38. Council on Environmental Quality, 1970 p. 70.

39. Ibid.

40. Council on Environmental Quality, 1971, pp. 107 et seq.

41. I. Michelson and B. Tourin, "Comparative Methods for Studying the Costs of Controlling Air Pollution," in *Public Health Reports* 81, 1966, pp 501–511.

42. Council on Environmental Quality, 1970, p. 72.

43. Ibid., p. 104.

Chapter IV

The Invisible Hand and Other Causes

Don't fell the tree that gives you shade.

Egyptian proverb

Modern economies are characterized by a great number of economic subsystems. In the United States, for instance, there are about fifty-seven million households and eleven million enterprises. A hundred thousand or more goods are demanded and produced. Each individual household and each enterprise independently pursues its own plans and its own goals. How can the interactions of such a complex system of individual organizational units be directed? How can one ensure that this economic anthill does not disintegrate and end in chaos? Two hundred years ago the English economist Adam Smith, the father of economic thought, believed he had found the answer.

> As every individual . . . endeavors as much as he can both to employ his capital in the support of domestic industry, and so to direct the industry that its produce may be of the greatest value; every individual necessarily labors to render the annual revenue of the society as great as he can . . . he intends only his own gain, and he is in this . . . led by an invisible hand to promote an end which was no part of his intention.[1]

Has this invisible hand not caused very visible results, such as the smog of Los Angeles and the fish dying in the rivers, the chemical clouds of industrial centers and the fumes from the automobile exhausts, the pollution of the oceans and the millions of tons of known and still unkown pollutants, and the nondegradable pesticides with their long-term effects about which we know nothing? Has it not wrought changes in the elementary consumer goods necessary to human life, such as air and water, and has it not disfigured our landscape and destroyed nature?

Marxist critics of the market system have emphasized these points. According to them, the capitalist system is the major cause of environmental pollution. "The environmental catastrophe is only one of many aspects of the basic contradictions in the capitalistic mode of production."[2] Or, "The extensive threat to and destruction of today's environment is a result of private ownership's centuries-long domination of the means of production, and especially a result of monopolistic capital in the twentieth century."[3] Environmental pollution, according to this thesis, is attributable to the capitalist mode of production with its core of private ownership and the profit motive. The very pursuit of individual economic interests—the leitmotif of the market economy—causes the damages. Self-interest begets pollution, or, in the terms of a German Marxist economist, "through the profit motive to environmental disruption."[4]

LAKE BALKHASH IN KAZAKHSTAN

Is this thesis correct? The fact that environmental pollution exists in market economies cannot be glossed over. But is the market system *the* cause of pollution? Is it a necessary prerequisite for such degradation, and does this phenomenon occur only in market economies? If so, then pollution should not occur in socialist economies.

A Marxist considers these questions to be nothing more than a smokescreen strategy for the justification of the market system. An objective analysis of the available information, however, indicates that environmental pollution is also a problem in the socialist economies.[5]

In setting prices, the socialist systems accord priority to goals in specific sectors, especially with respect to capital formation and resource extraction. The environmental effects of activities in the individual branches of the economy are not registered in, for example, the benefit indices. Nor are the prices for resources set with an eye to conserving these resources for the future. Concern has also been expressed that resources are being tapped too intensively. One explanation is that access to land is free. New land does not cost anything, so there is no incentive to exploit less rich veins in old mines. Since no importance is accorded to the price of land for a new site, there is a tendency to change sooner to a new mine than in an economic system where the price of land must be included in the calculations. The recently introduced land rent in the USSR for mining is not graded according to expected yields and therefore does not resolve the problem of misallocation, since the rate is uniform on all locations.

The goal of economic growth is accorded high priority in socialist systems just as in capitalist systems. There are many examples in which production managers opted to pay fines for environmental pollution in order to receive the significantly higher premiums awarded for surpassing the plan.[6] The heavy pressure to fulfill production quotas makes managers unwilling to buy equipment for pollution abatement, because such outlays would drain funds from growth-related investments. The priority scale at the enterprise level simply reflects the hierarchy of goals in the overall economy; production and growth figures are the measures of success both for a plan and for a manager.

Nothing is being done in the Soviet Union at the present time to attribute environmental damages to individual polluters. Therefore, the socialist production managers, like their capitalist counterparts, have no incentive to reduce air and water pollution.

Cost-benefit analyses, on which economic decisions are based, do not include all factors in their calculations. Thus, the rerouting of the Amu and Syr Dara rivers will probably lead to the disappearance of Lake Aral by the year 2000. Cost-benefit analyses of this project accounted neither for aesthetic values nor for possible effects on the climate. Especially in the case of dams, land values are often set too low in cost-benefit analyses, perhaps intentionally in order to make a project appear more profitable.

A striking example is Lake Balkhash in Kazakhstan. Unfortunately, it is not an isolated case. In 1965 a large dam was proposed for the Ili near Kapchagaisk. This river supplies 75 percent of the water of the lake. The dam would significantly improve navigability, encourage development of the local fishing industry, and provide water for irrigation. After the plans were drawn up, doubts surfaced about the wisdom of the project for environmental reasons. The city of Balkhash, lying with its 77,000 inhabitants in the middle of a desert, is very dependent on the lake, which provides drinking water, fish, and recreation. Because of the desert environment, it was feared that the high rate of evaporation with little incoming water would cause the salinity of the lake to rise and adversely affect the drinking water supply of the city. It was further observed that the dam would interfere with the spring floods in the delta, thereby hampering agricultural production. None of these factors were included in the cost-benefit analysis.

Management specialists have long been aware that every organization develops its own life and pursues its own goals. The same is true of planning divisions. As it becomes clear that it was no longer possible to persuade people with promises of agricultural benefits, the sales pitch shifted to energy production, a point which had not even been mentioned in the original plan. This required that the dam be constructed higher. After the project was approved, a new assessment discovered that the increased productivity would recover invested capital only after four years, rather than the initially estimated year and a half. After the dam was completed, the mineral content of Lake Balkhash rose 8 percent by the first half of 1970, as had been predicted by critics. Clearly, the project's effect on the quality of the environment was not sufficiently taken into account in the decision-making process. And the problem is not a thing of the past.

Today's visitors to Eastern European industrial centers such as the Silesian coal area will observe that air and water pollution are sizable problems.

These examples should suffice to document the existence of environmental problems in socialist economies. A reader who remains unconvinced might lend more weight to a statement expressed by Erich Honnecker, Secretary of the Socialist Unity Party of the German Democratic Republic, at the Eighth Annual Party Conference: "Environmental protection, the struggle against air and water pollution, the reduction of industrial and traffic noise are problems whose significance will increase in the years to come."[7]

SYSTEM-INDEPENDENT CAUSES

The fact that environmental pollution is observed in both market and socialist economies seems to indicate that there are causes independent of the economic system which lead to environmental degradation. What are these factors?

Wastes are by-products of production; i.e., several goods result simultaneously from a single production process. Sometimes these goods can be used, such as gasoline and heating oil, which are joint products of the refinement of crude oil. In these cases, they create no problem other than cost allocation. Often, however, no use can be found for one (or more) of the by-products; it is simply not wanted. Cooling water, for example, fetches no price on the market, and it is disposed of in the cheapest possible manner.

Increasing industrialization is accompanied by significant rises in productivity, so that more can be manufactured with the available materials. In the United States the GNP has risen (in constant price terms) from $203.6 billion in 1929 to $821.1 billion in 1974—a fourfold increase. The growth in goods produced has been accompanied by an increase in by-products. It is estimated, for example, that the amount of sulfur dioxide discharged into the air in the United States from 1940 to 1970 rose from 22 million tons to 34 million tons; of carbon monoxide from 85 to 147 million tons; and of oxides of nitrogen from 7 to 23 million tons.[8]

Increasing industrialization is related to a restructuring of production activities. Industries producing chemicals, automobiles, and atomic energy are growing rapidly. New raw materials are discovered and exploited, new industries and methods of production are introduced, new goods are constantly being promoted, increasing the diversity of unwanted by-products and creating new types of wastes and pollutants.

Consumption, too, is a two-sided coin, with benefits accompanied by unwanted by-products: carbon monoxide and oxides of nitrogen from cars, sewage and full garbage cans from daily consumption, and dumping grounds as reminders of our discarded durable goods.

New sources of energy have been developed both for production and for consumption. The need for energy has risen more than production and income in the last hundred years. Since the majority of the best-known air pollutants (carbon monoxide, sulfur dioxide, hydrocarbons, and oxides of nitrogen) result from incomplete combustion processes during the production of energy, this factor acquires considerable importance. In addition, new sources of energy mean new wastes and new pollutants, like radioactive particles and immense amounts of cooling water which can cause significant temperature changes in all water systems.

The increase in population automatically means a rise in production and overall consumption even if per capita consumption remains constant. This is accompanied by more waste. Studies show, however, that the increase in population has not been the deciding factor in the snowballing wastes of the past few years. The per capita wastes have risen significantly. In the United States, for example, solid wastes rose from

2.75 pounds per day per person in 1920 to 5 pounds in 1970. This figure is expected to reach 8 pounds by 1980.[9]

The spatial concentration of the population and of industry in major urban areas, along with broad belts of settlements along coastlines and in river valleys cause a heavy concentration of pollution in a few areas. Growing agglomeration results in more serious environmental pollution in these regions.

Consumption patterns are a decisive factor contributing to pollution. The characteristics are well known: the rich supply of consumer goods, the possibility of replacing damaged durable consumer goods with new ones, the social custom of buying an object's most up-to-date model to substitute for a still usable older version, the freedom from upkeep provided by the seemingly endless supply of new goods, and the shortage of skilled labor, especially in the area of repairs. This society is a major cause of worsening environmental degradation. The influence on consumer sovereignty exerted by the "secret seducers" of Madison Avenue and the fabrication of needs are factors contributing to pollution. Not least among these societal causes is the lack of an environmental ethic and the distinct emphasis on the attainment of individual goals without consideration of social interests. In our presumed mastery of nature we once smiled condescendingly at old cultures and their conceptions of the environment, such as the "oneness with nature" of the Pueblo Indian cultures. But perhaps the values which emphasize environmental conservation by prohibiting its destruction play an important role in social preservation.

Our environment has limits to its assimilative capacity, but it is being burdened with increasing wastes and pollutants. When pollutants reach a certain level, the threshold of assimilative capacity is quickly crossed, and the regenerative process of nature can no longer decompose, neutralize, or eliminate the pollutants.

Problem awareness has increased. It has grown not only as a result of greater environmental damages; the environment itself has also achieved a new status. A growing GNP and more abundant private goods have shifted the environment higher on the value scale. Problem awareness is interconnected with wealth and level of development. This also holds true for socialist countries, since socialist writers explicitly recognize that such factors as industrial development and population density are at the root of the problem. For example, the East German economists Grundmann and Stabenow state:

> . . . the dimensions of the demands on nature and of the changes wrought by man in it, as compared with the dimensions of nature itself, its supply of oxygen, water, raw materials, land, forests, potential recreational areas, air, light, plants, and animals have grown considerably, especially in industrialized and in densely populated countries.[10]

That environmental pollution poses a problem in all types of economic systems does not permit the conclusion that the economic and social systems have no influence on the *extent* of environmental degradation. On the contrary, each system has features that regulate the use of the environment in one way or another. We now explore some of the variations.

THE ENVIRONMENT AS A FREE GOOD

Economic systems with decentralized control leave production and investment decisions to autonomous decentral units. The control over these units is guided by a price system, for example by a system of market prices which are determined by the supply and demand of the market. Prices function to indicate the shortage of a good. High prices—a severe shortage—stimulate production, and a relatively low price for a good causes production of that article to shrink. Thus, prices are information signals for decentralized units and simultaneously represent incentives to increase or reduce the production of certain goods.

The question every economist ultimately seeks to answer about the "right" allocative system cannot be pursued here. It is sufficient to note that decentralized control through a price system—for which increasing support is being found among economic theorists in socialist countries—has the advantages of delegating decision making to smaller economic units, of flexibility in production decisions, of inducing technical progress, and, finally, of minimizing the resources needed for the coordination of individual units.

For a variety of reasons, this allocative mechanism cannot be permitted to function uncontrolled.

First, the built-in tendency toward monopolization in economic systems must be limited by active policies promoting competition, antitrust legislation, and open markets. Second, the price mechanism also guides the use of resources, thereby determining the distribution of income and, consequently, of wealth. Policies must correct undesirable patterns of distribution. Third, there are sociopolitical reasons for modifying the allocative mechanism. Where undesirable results occur, an in agricultural policies or in sectors having a strong effect on employment levels, the price system is replaced by other organizational and guiding mechanisms. Fourth, the price mechanism does not regulate the supply of public goods such as health care, education, and the judicial system. The supply of these goods is determined by a political decision-making process. Fifth, greater consumer sovereignty may also be desirable. The decentralized system of control rests on the thesis that through adjustments in supply and demand, the price mechanism directs production toward scarce goods, toward those goods with the higher prices. In order for this to be the case, however, demand must not be lured by advertising to those areas in which enterprises just happen to be producing goods. Business should produce and consumers should demand. Lastly, the observed degradation of the environment creates yet another need for modification. Before examining this aspect, however, the laws which determine thc uses of the environment in a market economy must be clarified.

In a decentrally controlled system, the use of resources, i.e., factors of production, is guided by the price of these factors, If there is a high demand in one sector of the economy for a particular factor of production, this factor's price will rise in the sector until an equilibrium is reestablished vis-à-vis the demand in other sectors. This is effected through the price mechanism, which guides the influx of this resource into one sector and its exodus from others. The factor that is no longer in as much demand in one sector will fall in price there, and the factor will be directed elsewhere. In this way, the price regulates the long-term allocation of production factors to the individual economic branches.

Resources which are needed for production but for which no price is demanded can be employed without restriction. Since a zero price creates no incentive for the demander to use the resource sparingly, it should come as no surprise that the demand will be excessive. The environment has long been considered this kind of free good, and as a result it has been overused. The environmental degradation found in market economies is causually related to the concept of the environment as a good with zero price value.

As long as the demand for such an unpriced good remains limited in relation to the supply, there is no problem. However, when the demand for the good rises because of population growth or because of the popularity of this "free" good, the quality of the good falls. There are countless examples of this phenomenon. The common—the communal pasture of English villages during the Middle Ages—was hopelessly overgrazed. The fish stocks of the world oceans, another free good, are being mercilessly reduced by improved fishing methods. First the blue whales were decimated with the use of larger boats and expanded fleets. When these largest and most profitable whales had been killed off, hunters moved on to catch the fin whales. When these have been depleted by excessive hunting, the sei whale and finally the sperm whale will be the next to go.[11] The point here is not to be sentimental, but to indicate the consequences of treating the environment as a free good which everybody can use in any way they please.

The lot of the blue whale and fish stocks is shared by other free goods. Since water can be used for any purpose by individuals on whose property it flows, it is badly overused as a dump for the unwanted by-products of production and consumption. This is true both for rivers such as the Rhine and for other bodies of water such as Lake Erie, the Caspian Sea, and the world's oceans. Man abuses free air just as carelessly when disposing of his countless wastes. And the polluters are never billed.

We are too accustomed to considering air and water to be abundantly available. The idea that air might be a scarce resource is totally alien to us. But the supply of these goods is limited and the demand is soaring. More oxygen is consumed in the United States than is produced by the vegetation in the country. And the earth's supply of oxygen is limited. "At an altitude of about 3,500 meters above sea level, car engines do not function well—and the same is true of humans."[12] The oxygen in our air is not just simply present, it must be produced. Forests and other terrestrial vegetation manufacture some oxygen, but the major source is the sea, where phytoplankton serves as the world's oxygen factory, providing 70 percent of its supply.

The overexploitation of the environment is aggravated by the pressure of increasing industrialization and growing populations. Examples of environmental pollution can also be found in small populations and in pre-industrial times, but they are then usually connected with a few areas of higher population density. Basically, most areas in pre-industrial times enjoyed an environment that was an abundantly available free good whose limits had not yet been discovered. In this situation it was possible to apply a zero price to this good. However, the environment has become a scarce good. The demand for environmental services has risen so sharply that the quality of this public good is bound to fall if it continues to be used free of charge.

We maintain that the treatment of the environment as a zero price good, its general cost-free use as a receptacle for wastes and pollutants, lead to overexploitation.

THE COMPETING USES

The use of the environment for production at zero price is problematical because various uses compete with one another. If an environmental medium were used only by a single agent for a single purpose there would be no such thing as environmental pollution. If water were used for cooling and for nothing else, no damages would arise. Unfortunately, we also need drinking water. If air were only employed as a receptacle for poisonous wastes, we would not have to worry about this environmental medium. Unfortunately, we must also breathe this air.

The concept of competing uses encompasses two different cases which overlap in reality, but which must be considered separately for analytical purposes.

Case 1: Single purpose competition. Suppose that a particular environmental good, a lake, for example, is employed for a single purpose, say, swimming. (This is, of course, a case seldom encountered in reality since in principle alternative uses are always possible.) If more and more people swim in the lake, then after a certain point the activities of person A will affect those of B. The limit of the environmental good will then have been reached. Or, imagine that the water in a river is used for the sole purpose of cooling wastes from power plants. Competing uses arise when a cooler upstream affects the process of a cooler downstream by returning warm water to the river. The same problem occurs in the case of communal land on which only a specific number of cattle can be allowed to graze without reducing the quality of the environmental good. Another example is a natural park. The popular ones in the United States are frequented by so many visitors that control measures are being introduced, such as limiting the number of people permitted daily on particular trails so as to protect the free good. The reasons for this type of competing use lie primarily in the growth of the population and in its spatial concentration. Changing preferences also play a role. The resulting burden on public goods and the overestimation of their capacity reflect this aspect of competing uses.

Case 2: Multipurpose competition. The other, more common case of competing uses occurs when an environmental service is utilized not for a single purpose but for many purposes. In reality the problem of the coolers depicted in Case 1 is complicated by water's alternative roles in the disposal of toxides, acids, phenols, and other pollutants; for the production of beer; for irrigation; and last but not least, for drinking. The number of conflicts multiplies extraordinarily rapidly in the case of water, for which there are so many simultaneous functions. Examples could be listed for other media, too. Through dust formation and general diffusion in the air, phosphates affect the use of land in the regions where they are extracted. Cement works influence recreational areas. The use of the environment as a wastebasket for industrial production also degrades the quality of the air we breathe.

The competing uses of the multipurpose type occur not only between two different production activities, nor simply between production and consumption. They also appear between different consumption activities. The use of a lake for motorboating reduces the fun of swimming. And swimming pollutes drinking water. It is the task of an economic system's control mechanisms to resolve this problem. If economic theory has taught us anything, it is that society must choose between different possible uses for resources, deciding which uses are most important, and allocating scarce resources to the individual competing uses. In other words, society must economize these resources, and the price mechanism is one possible mode of economizing.

But if the environment is treated as a free good, if it has a zero price, then one cannot expect the price mechanism and the market system to allocate the environmental services to the competing uses. The earlier definition of the environment as a public consumer good which is available to everyone, and whose use by individual A does not affect the use by individual B is clearly unacceptable in view of the discussion concerning competing uses of the environment. When such a public consumer good is utilized by many people the character of the good is lost. The good can still be employed by all, but its quality is degraded. Exhausts from cars and industrial plants can be emitted into the air, but this pollutes the air we breathe. Public consumer goods, therefore, can be enjoyed equally by A and B only until the limits of the good's capacity have been reached. Beyond this point, the character of the public good is altered. And in the case of the environment, this limit has long since been transgressed.

As a rule, we assume that these competing uses are freely interchangeable. A beautiful landscape can be used either as a natural preserve or as a mining site; a lake can be a source of drinking water or a dump for wastes. A forest will serve as a local recreational area or as a source for lumber. Unfortunately, however, these uses are not always freely interchangeable once the mode of use has been established. An area designated as a recreational park can later be mined. But the reverse is not a two-way street. When the area has been mined, the option to use it later as a park is, for all practical purposes, eliminated. Krutilla[13] refers to this as *asymmetry of options*. There are countless examples of this problem and any decision involving competing uses must carefully consider it.

In the examples above we have concentrated on negative effects, whereby one activity is adversely affected by another. But positive effects occur, too, as when forests benefit from the absorptive capacity of the soil which insures sufficient ground water, or when the photosynthesis of plants regenerates the air by removing the nitrogen. In these cases, the ground water or the atmosphere can be used for different purposes simultaneously. Unfortunately, such harmony of uses is not the rule for an environmental medium. It is the conflicting uses that are common and which create the problems of environmental pollution.

Up to this time, economic theory approached the problem of environmental pollution differently. It has not spoken of competing uses, but rather of externalities. An externality is when one economic activity is affected by another. There are two types. The first occurs in market processes in which the economic activities in question influence each other. Producer A lowers his price when competitor B lowers his, or when a substitute product competes for the market. The second type refers to technological effects goods have on each other as a result of the interdependence of economic activities through environmental systems. The air, for example, is the medium which transports the particles from phosphate mines to agricultural land; water is the medium which delivers chemical wastes into the water sources of breweries. This type of externality is at the root of the environmental problem.

The concepts of competing uses and negative externalities refer to the same empirical phenomenon, but from different points of view. The concept of competing uses starts from the environmental goods and discusses alternative uses to which they might be put. The concept of externalities, on the other hand, starts from an economic activity and examines how it affects other economic activities. The effect occurs, without being explicitly stated, through environmental goods. One can say that competing uses are the causes of externalities, and that externalities are the results of the unresolved competition between uses. Both approaches attempt to explain environmental pollution.

THE FLAW IN THE MARKET ECONOMY

Competing uses require a society to choose one use of an environmental medium at the cost of another use it might have. These allocative decisions must consider the alternative, or opportunity costs.

Opportunity costs refer to the benefit lost when the use of a resource for purpose A (opportunity A) precludes alternative use B (opportunity B). If water is used for the disposal of waste products from the paper mill, then the opportunity cost is the forfeited use of the water for beer production. If the water can be purified, the opportunity cost can be measured by the costs of purification. If that water had been used originally for drinking, the alternative cost after dumping wastes into it is the forfeiture of drinking water. The opportunity cost of phosphate mining is the decrease in productivity of the neighboring fields. The alternative cost of using the air to dispose of pollutants is the danger caused to health.

The problem of competing uses has not been resolved because the private economic accounts of the producer do not take into consideration the opportunity costs to society. Economic accounts include only the internal costs of each economic unit, but not the costs incurred to other economic units by externalities. Nor does this system of accounting take into consideration the costs to the consumer or to society at large. Social costs refer to those generated by both the individual producing economic unit, and, by virture of externalities, by other economic units (e.g., society). The problem, then, is that the use of the environment at zero price leads to a *discrepancy between private and social costs*.

A. C. Pigou devoted a chapter to this problem as early as 1929 entitled "Divergences Between Marginal Social Net Product and Marginal Private Net Product" in his *Economics of Welfare*.[14] He wrote:

> . . . uncompensated services are rendered when resources are invested in private parks in cities; . . . these, even though the public is not admitted to them, improve the air of the neighborhood. . . . It is true, . . . of resources devoted to afforestation, since the beneficial effect on climate often extends beyond the borders of the estates owned by the person responsible

> for the forest. . . . It is true of resources devoted to the prevention of smoke from factory chimneys: for this smoke in large towns inflicts a heavy uncharged loss on the community, in injury to buildings and vegetables, expenses for washing clothes and cleaning rooms, expenses for the provision of extra artificial light, and in many other ways.[15]

Further,

> . . . [incidental uncharged disservices are rendered] when the owner of a site in a residential quarter of a city builds a factory there and so destroys a great part of the amenities of the neighboring sites; or, in a lesser degree, when he uses his site in such a way as to spoil the lighting of the houses opposite.[16]

The discrepancy between private and social costs means that the prices of goods do not include all the social costs of production. For example, the goods which degrade the environment by emitting pollutants and which adversely affect the medium for other possible uses, do not reflect in their prices the damages and opportunity costs that their production entails. The costs of these goods are miscalculated; they are underpriced. What are the consequences?

Suppose we have two products whose production processes cause different degrees of damage to the environment—one is a polluting product and the other is environmentally sound. Take, for example, protein produced from petroleum by polluting processes. This is to be marketed alongside traditional steaks that are produced without environmental side effects. If no price is charged for environmental pollution, then the price of the polluting product, the "petroleum steak," does not include the opportunity costs of the damage done to the environment. The production costs are thus artificially low, allowing the price of the polluting product to be set too low vis-à-vis the true total costs. As a result, the demand for the polluting good will be higher and more of the product will be produced than if the environmental costs were reflected in the price.

This example also reminds us that the relative price—and not the absolute price—influences the demand. At a given level of income for a household the demand for the more expensive product will be lower. If the cost of using the environment is taken into account in the price of the product, the environmentally degrading product will become more expensive and the environmentally sound product will become more attractive in relation to its competitor. This would cause a readjustment of demand in favor of the environmentally sound product.

Yet there is no charge for the use of the environment, and not only are the media overexploited, there is also no trend to discourage production of environmentally damaging goods. The distortion of the relative prices causes in turn a distortion of production in favor of polluting products. Specifically, this means that too many factors of production are allocated to the pollution-intensive sector. Consequently, factor allocation is biased against the environment and in favor of pollution. Sector structure, too, is biased in favor of pollution-intensive activities. The zero price set on the environment distorts factor allocation and sector structure by acting as a sector-specific subsidy or as an input restraint. The same problem can also arise in a socialist economy when the environment is accorded a price of zero and when a discrepancy exists between private and social costs.

The cause of environmental degradation in a decentralized system lies in the fact that producers can generate costs which they do not have to bear and which are not attributed to their products. It is also rooted in the fact that households create costs for other households through their use of the environment as a receptacle for wastes, and that they are not billed for these costs. The accounting system is responsible for these problems. The flaw of the market economy is in the mode of accounting, which is based on individual economic units, and the possiblity of shifting the burden of the costs to society as a whole.

The obvious conclusion to be drawn is that a change is needed to establish a balance between the relative prices of polluting and environmentally sound products. This is being suggested in both market and socialist economies.

> Accounting for the costs of the environment and for the demands made on territorial resources will probably lead to significant changes in prices (especially in the chemical industry), and thereby to changes in the measures of efficiency for individual enterprises and branches of industry.[17]

If it were possible to force business to include the social costs of environmental pollution in the prices of its products, the present tendency to favor the production of polluting products would disappear.

Pigou concludes:

> No "invisible hand" can be relied on to produce a good arrangement of the whole from a combination of separate treatments of the parts. It is, therefore, necessary that an authority of wider reach should intervene and

> should tackle the collective problems of beauty, of air, and of light, as those other collective problems of gas and water have been tackled.[18]

Since Pigou's time we have had some experience with "authorities of wider reach" and the question now is: to what extent does the correction of market failures through government planning lead to planning and bureaucratic failures? We will discuss the role of economic policy decisions in Chapter VII and options to reduce the discrepancy between private and social costs will be examined in Chapter IX.

THE REASONS FOR ZERO PRICE

Why has no price for environmental goods developed in the past as it has for other goods? This question is worth investigating because it uncovers the cause of the flaw in the market economy and also discloses the difficulties involved in correcting the problem.

The reason that there is no price attached to the use of the environment is that private ownership of environmental goods cannot be defined in the same legal terms that are applied to conventional private goods. Light and air cannot be packed into bags like potatoes and allocated to individuals. The possibility of turning environmental goods into private goods might well strike the fancy of experts in capital formation, and finance ministers might rejoice at the potential sources of revenue. But apart from the macabre aspects of such thoughts, it is not technically possible to convert environmental goods into private goods so that their use by individual A does not have consequences for individual B. The environmental media cannot be arbitrarily divided into private property. And, as in the example of the water running through an individual's parcel of land illustrates, private ownership does not remove the problem of externalities as long as technological links exist between the individual environmental units.

The solution to the problem of environmental pollution does not lie in the introduction of private property laws in the domain. On the other hand, this also shows that the absence of private property—as demanded by Marxists—is no guarantee for the absence of environmental degradation. In fact, the fate of the common, cited earlier, is the best example of how publicly owned goods can be destroyed through overexploitation.

The fact that it is technically impossible or politically undesirable to turn the environmental media into private goods does not mean that no property laws can be designed for the environment. Property involves a great variety of societal rules for the use of goods. Chapter IX treats this subject in detail.

Another related approach interprets environmental media as public goods which, because they are public, often possess no price. As explained earlier, a characteristic of public goods is that they can be used by all individuals simultaneously. A national park is available to many visitors. Nobody can be excluded from using public goods. The environment is a public good because it is impossible or undesirable to assign a price to it.

Assume that an enterprise could produce fresh air. Air that is purified for customers is simultaneously purified for the public at large. Since the firm cannot exclude free-riders, it has no incentive to produce this public good. And even assuming that the private firm could somehow exclude individuals who are unwilling or unable to pay for the product, this exclusion would not be considered socially desirable.

The nature of the environment as a public good and the impossibility of making it private are related reasons for the nonexistence of a price. There arguments can explain why no price was demanded in the past for goods which were originally abundant. And the arguments indicate why the use of these goods was not included in business accounts. These reasons also clarify why the flaw in the market system exists and they indicate the difficulties which must be overcome in order to correct it. The problems raised in this chapter are the groundwork for the possible measures explored in Chapter IX.

NOTES

1. Adam Smith, *An Inquiry into the Nature and Causes of the Wealth of Nations,* 1776, Book 4, Chapter 2.
2. G. Kade, "Durch das Profitmotiv in die Katastrophe," *Wirtschaftswoche* 40: p. 44 (1971).
3. S. Grundmann and E. Stabenow, "Beziehungen von Mensch und Umwelt," *Wirtschaftswissenschaft* 12, December 1971, p. 1178.
4. G. Kade, op. cit.
5. M. I. Goldman, "Externalities and the Race for Economic Growth in the USSR: Will the Environment Ever Win?" *Journal of Political Economy* 8, 1972: 314–327; and *The Spoils of Progress: Environmental Polllution in the Soviet Union,* Cambridge, Mass., 1972.
6. See, for example, *Der Spiegel* 9, 1973: 50.

7. S. Grundmann and E. Stabenow, op. cit., p. 1774
8. Council on Environmental Quality, 1972, p. 6.
9. Ibid., 1970, p. 106.
10. S. Grundmann and E. Stabenow, op. cit., p. 1776.
11. D. H. Meadows et al., *The Limits to Growth,* London, 1972, p. 151.
12. D. Widener, *Kein Platz fuer Menschen: Der programmierte Selbstmord,* Frankfurt, 1972, p. 49.
13. A. C. Fisher, J. V. Krutilla, and C. J. Cichetti, "The Economics of Environmental Preservation: A Theoretical and Empirical Analysis." *American Economic Review* 62: p. 605–519 (1972).
14. A Pigou, *The Economics of Welfare,* London, 1929.
15. Ibid., p. 184.
16. Ibid., p. 185.
17. S. Grundmann and E. Stabenow, op. cit., p. 1784.
18. A. Pigou, op. cit., p. 195.

Chapter V

The Hitch to Economic Growth

When you bend in one direction, you necessarily turn your backside to the other.
Chinese Proverb

Ecologists consider the growth fetish of Western industrial societies to be another important cause for environmental pollution. Growth served as the political slogan in the elections of the last decade. Growth was the motto of the European Economic Community, of chambers of commerce, of unions, and of politicians. And growth was the goal to which official economic policy was solemnly committed.

Significant growth has indeed been registered for the period between the end of World War II and the advent of the energy crisis. The United States increased its GNP from $355.2 billion in 1945 to $821.1 billion in 1974 (in constant 1958 dollars).[1] The FRG quadrupled its GNP in the twenty-one years between 1950 and 1971.[2] Other nations have recorded even more spectacular results. In the twelve years from 1958 to 1970 Japan's real GNP quadrupled.[3] Socialist countries, too, were also characterized by high economic growth. For example, the net material product in Yugoslavia quadrupled between 1960 and 1969, admittedly in current prices.[4]

SOCIETY FOOTS THE BILL

This growth, according to ecologists and critics of growth policy, was achieved at the expense of the environment. It was possible only because the wastes and pollutants resulting from increased growth and consumption were dumped into the environment without regard for their effects, and because economic accounting systems did not calculate the destruction of the environment, the adverse effects on the quality of environmental services, the degradation of the air, and the contamination of the water. Nor did growth-oriented accounting systems recognize the reduction in value of many environmental goods. Moreover, the social costs of growth were not included in the economic calculations.[5]

Take a single example, say, the use of chemicals in agriculture. The amount of herbicides employed in the Federal Republic of Germany rose by 400 percent in the seven years between 1962 and 1969. Fungicides increased over 200 percent in the same time span.[6] Diagram 4 illustrates the development of agricultural production in the United States according to acreage cultivated and nitrogen fertilizers employed.

The relatively low level of growth in production is accompanied by a sevenfold increase in nitrogen usage. The simultaneous decrease in acreage under cultivation indicates that large amounts of artificial fertilizers were concentrated on less land.

Diagram 4.[7] Ratio of nitrogen fertilizers applied to acreage cultivated in the United States 1945–1970

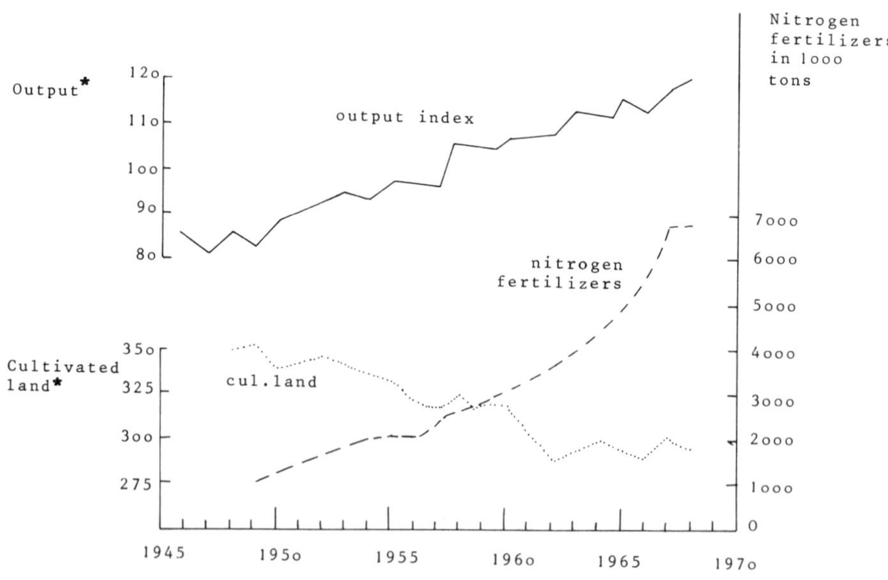

*The agricultural output is measured according to an index calculated by the Department of Agriculture.

Source: U.S. Department of Agriculture, *Agricultural Statistics,* 1967, pp. 531, 544, 583; 1970, pp. 444, 454, 481.

Nitrogen fertilizers are necessary because the intensive use of land makes it impossible for the bacteria which decompose the humus to release enough nitrates. Nitrates are absorbed by the plants through their roots and are then turned into protein. Nitrogen fertilizers can therefore be considered a substitute for the production factor "land." Production can be significantly increased through these artificial fertilizers.

Unfortunately, undesirable effects accompany the application of these fertilizers. The reduced humus content of the soil, which results from the intensive cultivation, makes the plant roots incapable of absorbing all the additional fertilizer in the less porous earth. A large part of these fertilizers is thus washed away, eventually reaching the rivers, ground water, and lakes. Nitrates stimulate the growth of algae, and the resulting increase in both live and dead algae raises the oxygen demand of the water, whereupon the biological purification capacity of the water can be destroyed. Presumably, nitrates have no negative effects on humans, but nitrates can be turned into nitrites by intestinal bacteria, especially in children. Nitrites attach themselves to hemoglobin, the oxygen transporter in human blood, and thereby reduce the oxygen supply. Damage to health can result.

Ecologists do well to point to such examples. Chapter III cites many in which the repercussions on humans range from effects on the plant and animal world to possible changes in the atmosphere and in other essential systems of nature. These examples support the thesis of ecologists that we have achieved growth at the expense of the environment.

This thesis cannot be denied. We must ask ourselves what the consequences are. Ecologists draw the conclusion that the goal of economic growth should be abandoned immediately.

PROS AND CONS OF A ZERO-GROWTH ECONOMY

A halt to economic growth involves many highly unpleasant effects. Economic growth means an increase in goods produced. Maybe this increase is no longer so important for the industrialized nations, but it is an absolute necessity for the countries of the Third World. A stop to economic growth in the Third World would lead to chaos in those countries where medical care has raised the life expectancy of the population, where birth rates continue to multiply, where industrial development is only in its initial stages, where poverty is extensive, where essential goods are still not available, and where the per capita income is under $200 a *year*. For these countries the increase in goods is a must.

Zero growth also raises a number of problems for developed economies. The supply of public goods in the important areas of education, housing, medical care, and hospitals would be restricted. It cannot be denied that there is still a significant shortage in these sectors today.

In the past, growth has permitted mitigation of distributional conflict and the reduction of social tension by allowing the absolute improvement in the situation of each individual. A growing economy also makes relative redistribution easier to accomplish—if it were to be seriously attempted—than would be the case in a static economy. The example of Great Britain (before the discovery of North Sea oil) seems to demonstrate that distributional conflicts become more severe under stagnating or declining economic conditions, and that with greater stress on the equity goal, economic efficiency and economic growth are negatively affected. But no one can claim that environmental quality will improve in such a situation. On the contrary, low growth, recession, and the accompanying lack of public funds extinguish incentives for enforcing environmental policy.

Finally, growth permits more free time through a reduction in working hours. And if one considers income security, job security, and security in the years after retirement, then growth clearly allows—in addition to the greater amount of goods—more flexibility, freedom, and security than an economy at a standstill.

Even the readers who do not agree with the above view must admit that a halt to economic growth is totally incapable of removing environmental pollution. A halt to growth would leave the economic activities at their present level with no change in the environmental degradation to be seen. A halt does not even slow the further deterioration of environmental quality. Since pollutants continue to be disposed of and accumulated in the environment, environmental quality can fall significantly in spite of zero economic growth.

There is another reason why economic growth and environmental protection are not necessarily incompatible. Environmental protection requires new environmentally sound technology and major investments to eliminate wastes from consumption and production. It also requires the recycling of material. Environmental protection thus entails many growth impulses which can draw the economy into a new stage of development.

Growth per se does not aggravate the conflict in goals between environmental quality and the production of goods. Rather, the problem lies in the way growth has been achieved up to now. The disproportionate amount of polluting products over the environmentally-positive ones is one major contributor to the conflict. And as emphasized in Chapter IV, the free use of the public good environment is another.

The conclusions we draw from these observations reiterate our earlier statements. The environment can no longer be used as a free good for all possible competing purposes. The competing uses must be evaluated and the use of the environment must be accorded to the purposes with highest priority. The evaluation must also compare the environmental services to private goods. This means that not only the private goods but also the public goods (in this case, environmental services) are to be included in measurements of well-being. Such a measurement also reveals that growth has an instrumental character; it is a means to raise the level of well-being, not a goal in itself.

This two-dimensional measure of welfare—goods and environmental quality—also shows that in past years the growth rate of well-being was significantly less impressive than that of the goods produced. And it shows that well-being in the future will, at least in the rich economies, be more compressed, like the blades of closing scissors.

First, the increase of the goods produced through trivial variations of existing goods, through new packaging and coloring, through new, up-to-date models, does have a purpose, but achieving greater well-being through additional goods is becoming problematic. The ultramodern family with one house, one plane, two boats, three cars and four TVs as described by Mishan is an apt caricature of the problem of super-consumerism.

Secondly, the expansion of the goods produced is accompanied by ever increasing pollution and deterioration of environmental services. If the environment is the determining factor for well-being, then the increasing goods and the related pollution actually erode the level of well-being. In addition, the greater amount of leisure time afforded by economic growth raises the demand for environmental services and thereby the value accorded to them.

A comparison of values accorded to private goods and environmental services over time indicates that the value of the former has fallen as the supply has increased, while the latter has become more valuable. How much was a loaf of bread worth to a peasant in the time of Luther? And how much is it worth to a New Yorker today? On the other hand, how much did the peasant have to pay for fresh air in Luther's time?

nothing. And yet how much would we be prepared to pay today in many cities?

The environmental pollution which results from the kind of economic growth we have been experiencing leads to the conclusion that it is necessary to change the values accorded to private and public goods. The zero price placed on environmental services up to now must be replaced by a *positive* value. This involves a revaluation of the goods produced, and it requires a restructuring of production.

THE INS AND OUTS OF THE GNP

Economic growth is measured in terms of the Gross National Product (GNP); that is, according to the value of the goods and services produced each year. This figure totaled $1,691.6 billion for the United States in 1976.[8] After the capital consumption allowance has been subtracted to account for the wear on machinery, the GNP becomes the Net National Product (NNP). GNP and NNP are interpreted as indicators of the economy's productive capacity, or as a gauge of well-being.

An indicator intended originally only as a gauge of an important overall goal develops its own life, becomes as glamorous as Marilyn Monroe, captures the headlines, is courted by statisticians and politicians, and eventually overshadows the very goal it is merely supposed to describe. Shortly after it is introduced, only the indicator is "traded" in the news media. The end to which the indicator was once attached subsequently gets left out of the picture.

But the GNP is an insufficient indicator of well-being for a number of reasons.

1. The national product is presently measured according to the costs incurred in the production of goods, specifically the costs of the factors of production. For this reason, it has been suggested that this measure be called the Gross National Cost rather than the Gross National Product.[9] There are two other concepts which should be included in the definition of the national product. One is the goods concept, which provides information as to the amount of goods produced within a period. These goods should be measured by the value accorded to them by consumers rather than by the costs of production. The benefit concept, secondly, indicates the benefits of the goods present (not identical to those produced) in a period.

2. Public goods are insufficiently included in this measure of well-

being. As in the case of private goods, only the costs of production are represented. The national product measures only the government expenditures for education, not the benefits of the product "education"; only the expenditures for public administration, not the benefits accruing from it are included in the calculations. Also, national accounts record public goods only for the period in which they are produced. In the phase during which they are used, public goods are represented only through the costs of maintenance or service. If one wants a measure of well-being, then the benefits from these goods during each period have to be included, in effect combining the benefits from goods produced in the given period and those presently available and usable goods produced during previous periods.

3. In addition to public goods, many other durable goods provide benefits beyond the period in which they were produced. A beautiful tree, such as the ancient California sequoia, does not appear in the national product nor contribute to the measure of well-being until it is felled, sawed up, and sold. However, well-being does not depend on the amount of durable goods which are produced or sold in a period, but rather on how many can be used. Therefore, instead of accounting only for the costs of production, it is necessary to account for the benefit that flows from these durable goods, both for the means of production and for consumer goods. When the benefit is also included in the calculations, more weight will be given to the service life of these durables, particularly in the case of increasingly important consumer goods. The present process of accounting acknowledges the production of a good without concern for the length of its service life. Say, for example, that a certain good could be produced with a service life of either two or three years. If both involve the same costs of production, then both are calculated to have the same value in the present system. In fact, over the long run, the shorter-lived good will appear to be more valuable, because its replacement will be produced in two years' time, once again contributing to the GNP. However, in a national accounting system which measures the benefits of each period and thereby implicitly the service life of the good, the product with the longer life would be accorded higher value. At least this method would not honor the observable deterioration of product quality as an increase in well-being!

4. The amount of consumer goods, then, says little about the measure of well-being.[10] Well-being results from a sum of consumption acts of private and public goods, including the environment. As long as nothing

is known about the relationship between well-being and amounts consumed, little can be said about how the amount of goods produced affects the measure of well-being. Consequently, a pure multiplication of amounts of products does not necessarily mean an increase in well-being. Imagine that the supply of a certain product increases sharply. If the market system is functioning ideally, then the price of the product will sink in relation to that of other goods. Its weight in the measure of well-being will also fall. This price effect does not include the effects on the well-being of those goods which have no price, especially public goods.

5. In order to measure well-being, the "true" preferences[11] of the individual consumer must be determined. This is a problem. For one, preferences do not remain constant over time. They are determined by society in a social process, and individuals are influenced by advertising. But the basic difficulty is who shall determine the value of a private product or a public good? Should this be an omniscient being having insight into each individual? Or, in constrast to the fantasy of Orwell and the pretension of Hitler, should the valuation of private and public goods be based on the judgments of all the individuals in society? If so, how can individual values be aggregated into a common norm for society? The solution to this difficulty will contribute much to an improved system of social accounting. (See Chapter 7 for more detailed discussions on the determination and aggregation of individual values.)

6. Leisure time is not included in the Gross National Product, either.[12] Its increase in the past is not recorded in the measure of societal well-being, except perhaps indirectly through increased expenditures for travel and other recreational activities. Since the gauge called "national product" does not discriminate between costs and benefits, it does not permit the determination of an increase in the net benefit.

Should, then, expenditures serving only to maintain rather than raise the level of well-being be deducted from the national product? It is argued, for example, that military and police services only maintain a given level of internal and external security and that related expenditures should be deducted from the GNP. Capital consumption and capital expansion for the maintenance of growth would also have to be subtracted. The same holds true for the costs of adaptation which appear as a result of substantial sectoral shifts in the growth process, such as the costs of retraining and changing jobs. It is also questioned whether such expenditures as advertising budgets should be included in the GNP.

7. The national product reflects only monetary costs, not real costs.[13] It is problematic to account for such real costs quantitatively. The production factor of labor, for example, would have to be measured in terms of hours worked, but what would the resulting figure mean? Clearly, there is a significant qualitative difference between the thirty-ninth work hour of a laborer and the forty-eighth. Calculations are further complicated by the fact that the value of leisure time falls when free hours increase as work hours are reduced. Well-being is also influenced by the distribution of income and of wealth. In short, the all-encompassing term "Gross National Product" is sketchy at best. It spite of its many "ins," there are too many crucial "outs" to allow the GNP to be a sufficient indicator of well-being.[14]

Nordhaus and Tobin[15] have attempted to determine quantitatively some of the false estimates provided by the national economic account. They suggest the following amendments to the U.S. account for 1965: if leisure time and the work done on holidays were also included, then $925 billion would have to be added to the GNP of $560 billion for that year. Also unaccounted for are the services of private and public capital goods worth $80 billion. The following amounts would have to be subtracted: $95 billion for expenditures for internal and external security; $91 billion for capital consumption; $101 billion for capital expanding investments as a requisite of costs of economic growth; finally, $31 billion for environmental pollution.

THE ENVIRONMENT IN THE NATIONAL ACCOUNT

As the last item in the account presented by Tobin and Nordhaus indicates, one of the many problems involved in the definition of the national product is the environment. Environmental quality is treated in a very peculiar way in the present accounting system.

If growth in goods produced occurs, and this growth is only possible at the expense of environmental quality, then the deterioration is not reflected in any way in the national product. Paradoxically, a net investment which is made, say, to reduce the pollutants emitted by a plant, is entered as a "good," even if this investment only impedes potential environmental pollution. And this investment is still recorded as a positive contribution to the national product when in the same time period and in the same area the pollutants increase, either because of higher output in the plant or because other firms move to the area. Thus, the more we invest in the environment, the higher our standard of living, according to this system of accounting. But the investments could also mean that rapid deterioration in environmental quality is making the equipment imperative, and that the investments are not sufficient even to maintain a certain level of environmental quality. The national product includes the hours worked for environmental conservation, the government expenditures for environmental protection, as well as expenditures by households for such environmental equipment as air conditioners and antismog devices on cars.

To summarize, then, not only does the account ignore environmental deterioration, it may actually indicate an increase in well-being—because of a few expenditures for environmental protection—when the quality of the environment is falling substantially.[16] There is a major inconsistency in any system which accounts for the wear and tear on private capital goods, machines, buildings, and equipment, but not for the deterioration of the environment.

In order to resolve this problem, it would be necessary to ascertain the value of the deterioration of the environment to obtain the Net National Product. This would entail subtracting from the Gross National Product all the negative changes wrought on the environment within one economic period. This is easier said than done. How should the changes in the environment be measured, and how should the resulting figures (e.g., amount of pollutants) be assessed? This process assumes knowledge about societal values.[17] How can we compare the different societal perceptions of air quality at home and at work in Denver, Colorado, and Dallas, Texas? What values can we attribute to the pollution of drinking water, to the contamination of the Mississippi, and to the eutrophication of Lake Erie? And what relative values should be assigned to one environmental medium vis-à-vis another?

In the final analysis, the solution to the problem of determining a measure of well-being must lie in the subtraction of unwanted byproducts from the national product. In Chapter IX we shall discuss the suggestion that a tax or fee should be set for these products to reflect exactly the societal damages they incur. If such a tax could be determined, it would represent a negative value for the unwanted byproducts of consumption and production which could be subtracted from the national product.

We are a long way off from such a solution. Economists have suggested alternatives. One possibility is to include in the societal measure of well-being only those expenditures which raise the net well-being. Expenditures which serve solely to maintain the standard of living should be deducted. With reference to environmental quality, therefore, this would mean that expenditures for the maintenance rather than the enhancement of environmental quality would be subtracted from the GNP.

For the sake of consistency, then, all expenditures for the protection of a given standard of living should be deducted from the national product. The American economist Jaszi objects:

> . . . food expenditures defend against hunger, . . . clothing and housing expenditures defend against cold and rain, . . . medical expenditures defend against sickness, and religious outlays against the fires of hell.[19]

The "bookkeepers of the nation" are faced with a dilemma: which expenditures serve only to maintain a given level of well-being? According to which criteria should these differentiations be made objective and operational? Government statistics experts should be prepared to present the precise motivations behind the expenditures if they want to implement this differentiation.

Expenditures for environmental protection do not measure the deterioration of the environment; the latter can be significantly higher than the amount expended for pollution abatement. This would mean that the deduction should be even greater. Or the deterioration can be less than the investment for environmental protection. Then the Net National Product would be lowered more than it should be. It is possible to account for environmental degradation in this way only if expenditures for environmental protection equal the actual rate of deterioration within one economic period. Such a situation is not to be expected.

Finally, which level of well-being should be chosen as a base year for the standard to be maintained? Why should the air quality existing in 1931 be used instead of that, say, of 1950?

Although there is widespread dissatisfaction with the national product as a measure of welfare, and it is agreed that there is a pressing need to revise this indicator, the direction of the necessary changes remains unclear. Because it is presently impossible to make a comparative assessment of the goods represented in the GNP (or NNP) and the unrepresented, unwanted by-products, and because valuation of these by-products has been unsuccessful, the only solution lies in the use of two indicators, namely, the national product and a new indicator for environmental quality. The success of an economy and the well-being of its citizens should no longer be measured according to one index only, but according to two. Whether or not a given expenditure has raised the level of well-being will be decided by a comparison of the two indices. And this evaluation will be conducted either by the individuals for themselves, or by economists and public opinion for the population at large.

Using two indices for the two components of well-being forces both economists and citizens to divulge their preferences by comparing alternative levels of the Gross National Product with alternative levels of environmental quality, whereas a single index obscures the values that are more or less arbitrarily assigned to the envirnoment and the goods produced. The environment can be measured according to several indices which contain the various values. Indices have already been developed in the United States for air and water quality.[20] The figures for the expenditures by consumers, businesses, and government for the treatment and prevention of wastes must be accounted for in separate statistics, thus at least recording information on the efforts of an economy to conserve the environment.

The revision of the concept of the national product or the introduction of an indicator of environmental quality only alters the overall orientation of economic policy decisions. These changes provide no information about how the decisions made in the subsystems of the economy can be influenced to take environmental goals into consideration. Additional measures must be devised to transform the macroeconomic policy into guidelines and incentives for microeconomic units.

THE CONFLICTING GOALS OF ECONOMIC GROWTH AND ENVIRONMENTAL QUALITY

With a given level of resources, one cannot have endless amounts of both bread and bombs. A similar squeeze exists between environmental quality and economic growth or level of production. Diagram 5 illustrates this conflict for a very simple system.

The goods produced increase with the resources employed. This relationship, which is represented in the production function, is shown in the lower right quadrant. However, increases in resources used also entail rises in polluting wastes—in other words, environmental quality falls. This relation is illustrated by the curve in the upper left quadrant. Every unit of resource therefore corresponds to a particular amount of goods

Diagram 5. Environmental quality vs. economic growth

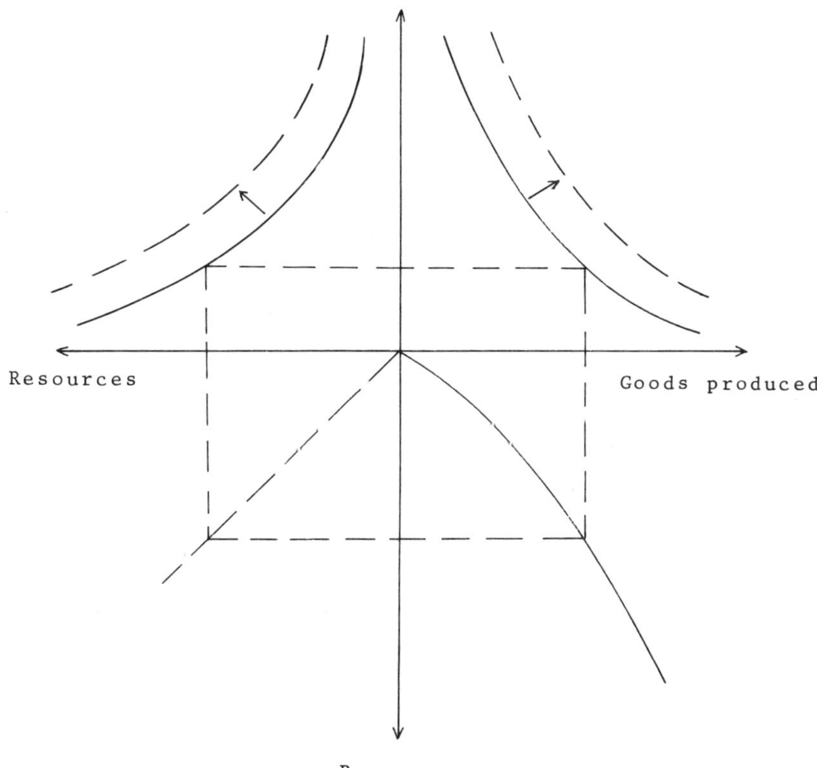

produced and a particular level of environmental quality. For any amount of resources, the combination of these two values can be found in the upper right quadrant. When these points are joined, they form the curve of the goal relationship between environmental quality and level of production. This curve has a negative slope. This means that to achieve a higher level of production, one must accept a lower level of environmental quality, and that a better environment can be obtained only if we are prepared to sacrifice a portion of possible production.

The conclusions of Diagram 5 are based on very simple premises, namely, the given production functions for goods and pollutants. If, for example, technological developments permitted a decrease in the amount of pollutants produced from a given amount of resources, the curve in the upper right quadrant would shift upward. This would represent a rise in environmental quality. If, instead of a single product, we were to observe a change in the production of many products, which is more realistic, then the shift of the curve in the upper left quadrant (i.e., mitigation of the conflict between the two goals) could also result from a restructuring of production in favor of less pollution-intensive products.

NOTES

1. Economic Report of the President, Washington, D.C. 1975, p. 250.

2. *Statistisches Jahrbuch für die Bundesrepublik Deutschland,* 1972, p. 514.

3. See *Statistisches Jahrbuch für die Bundesrepublik Deutschland* from 1965 to 1972.

4. United Nations, *Yearkbook of National Accounts Statistics* 1970, New York, 1972, p. 1378.

5. J. Mishan, *Technology and Growth: The Price We Pay,* New York, 1969.

6. Bundesminister des Innern, *Materialien zum Umweltprogramm der Bundesregierung,* 1971, p. 76.

7. B. Commoner, "The Environmental Cost of Economic Growth," in S. H. Schurr, *Energy, Economic Growth and the Environment,* Washington, 1972, p. 42.

8. This figure is in current prices. In 1970 prices it was $1,264.7 billion. OECD, Main Economic Indicators, Paris, 1977.

9. E. G. Dolan, *Tanstaafl* ("There ain't no such thing as a free lunch"), New York, 1969, p. 10. See also J. Joehr, "Bedrohte Umwelt: Die Nationaloekonomie vor neuen Aufgaben," in *Umweltschutz und Wirtschaftswachstum,* M. P. van Walterskirchen (ed.), Munich, 1972, p. 74 et seq.

10. B. S. Frey, *Umweltoekonomie,* Goettingen, 1972, p. 70.

11. Ibid., p. 53.

12. W. Nordhaus and J. Tobin, "Economic Growth," National Bureau of Economic Research, New York, 1972.

13. E F. Denison, "Welfare Measurement and the GNP," *Survey of Current Business,* January 1971.

14. This does not mean that an increase in the GNP is totally unreflected in societal well-being. R. Lampmann, for example, finds that a 26 percent increase in real GNP in the United States (1947–1962) meant an increase of 26 percent in per capita consumption, an increase in income security, and a decrease in poverty, thereby indicating a rise in the level of welfare. R. Lampmann, "Recent U.S. Economic Growth and the Gain in Human Welfare," in W. W. Heller, *Perspectives on Economic Growth,* New York, 1968, pp. 142–162.

15. W. Nordhaus and J. Tobin, op. cit.

16. The argument may be somewhat exaggerated. For instance, it can be shown that implementing environmental policy measures (an emissions tax) will, by making the pollution-intensive commodity more expensive, lead to a decline in the national income (defined in the traditional sense). See H. Siebert, "En-

vironmental Policy, Allocation of Resources, and Comparative Advantage," *Beiträge zur angewandten Wirtschaftsforschung,* Mannheim, 1977.

17. E.F. Denison, "Welfare Measurement," op. cit.

18. Ibid., p. 15; O. C. Herfindahl and A. V. Kneese, "Measuring Social and Economic Change: Benefits and Costs of Environmental Pollution," National Bureau of Economic Research, Conference on the Measurement of Economic and Social Performance (in preparation).

19. G. Jaszi, Comments on F. Thomas Juster, "A Framework for the Measurement of Economic and Social Performance," unpublished paper prepared for the National Bureau of Economic Research Conference on the Measurement of Economic and Social Performance, Princeton, N.J., November 4–6, 1971, p. 11.

20. Council for Environmental Quality, Annual Report 1972.

Chapter VI

Spaceship Earth

In the circumference of a circle the beginning and end are common.

Heraclitus

Kenneth Boulding[1] has compared the earth to a spaceship. Like a gigantic Apollo capsule, it moves through the universe with 3.6 billion people on board, encircled by thin protective layers, such as the magnetic field, the Van Allen belt, the ozone layer, and the atmosphere. The supplies are limited, but the passengers are multiplying at a terrifying rate. Production and consumption are causing ever more wastes and pollutants which, because they must be disposed of within the spaceship itself, deteriorate the quality of life. Waste production is endangering water supplies, and the air conditioner is not functioning correctly anymore so the air is becoming scarce and contaminated.

The spaceship earth is a closed system which receives nothing from the outside except energy from the sun. The majority of the energy and all food and goods must be produced with the resources available in the spaceship. The same system must also absorb the wastes generated by this consumption and production activity.

The fate of the people on board the spaceship can be compared to that of a modern Robinson Crusoe who, having missed the return "window" to his home planet, is stranded in the universe. Provided with a scarce amount of food and a limited supply of oxygen and water whose replenishment is impeded by the wastes he produces, this stranded astronaut is faced with the dilemma of which he will run out of first, food or air.[2]

ON-BOARD SUPPLIES ARE LIMITED

We intend to analyze this spaceship earth and present nine propositions which reflect the decisive elements of its closed system in economic terms. The economic policy of spaceship earth cannot be analyzed for one year alone. We need a long-term approach which incorporates the perspectives on future development, the processes which will evolve in the closed system, and the system's environment, all of which affect the living conditions on board. The interrelations between the environmental and economic systems discussed in Chapter II will now be placed in a time frame.

1. *Food and natural resources are limited.* Food and resources are the major preconditions for human life. However, the land available for food production is limited. Additional land can be placed under cultivation only with significant costs, say pessimistic experts. Worse, the growing population also demands more space for housing and transportation, which compete with agricultural needs for land.

Most of the resources for the production of nonfood items are limited, too. The raw-materials base and the related supply of energy limit the amount of goods that can be produced. Certain important resources are presented in Table 4 along with the maximum time they can be expected to last.

Pessimists argue that even if new types of raw materials are discovered, the effect they might have on the environment could be negative. The use of atomic energy will certainly increase the energy supply significantly, particularly when hydrogen can be substituted for scarce uranium as an input into the production process. But when so little is known about the possible negative effects of radioactive wastes on the

Table 4.[3] Present and Projected Supply and Demand for Selected Raw Materials

Raw material	Known reserves	Reserves at present level of consumption (years)	Expected rise in consumption (%)	Reserves at growing rate of consumption (years)
Aluminum	1,170 million tons	100	6.4	31
Coal	5 billion tons	2,300	4.1	111
Copper	308 million tons	36	4.6	21
Iron	100 billion tons	240	1.8	93
Natural gas	32,300 km3	38	4.7	22
Oil	72.5 km^3	31	3.9	20
Zinc	123 million tons	23	2.9	18

environment, about their storage, and about the extremely high demand for cooling water, the use of this resource has strict limits. By the same token, the discovery of minerals in the ocean, like magnesium nodules, will increase the quantity of known reserves, but we are not sure what environmental disruptions will result from their extraction.

2. *Only a limited number of resources can regenerate themselves.* Resources are the product of biological, chemical, physical, and geological processes in nature. The time required for these processes varies greatly. Geological time spans needed for the formation of coal and petroleum through sedimentation are beyond the horizons of human planning. The only regenerative raw materials, then, are those like plant and animal products which develop through biological processes. The regeneration of these natural raw materials can be described by biological growth functions. Regeneration initially proceeds rapidly, like the fish populations in a newly constructed reservoir, and slows after a certain point. The amount of the resource then sinks, eventually stabilizing at a level of equilibrium.

These growth processes change, however, when humans interfere significantly with the regenerative capacity of nature. Many fish populations have been so greatly reduced that regeneration to previous levels has not been possible. As a result, some species, such as whales, face extinction. The uncontrolled felling of forests has led to climatic changes and the calicification of the land so that no new forests can ever grow there again. The Mayas, for example, destroyed the primeval forests in Yucatán through their methods of agricultural production, and it is hypothesized that this was also the cause of their own destruction. Because all plants grow abundantly in this climate, fields can only be cleared by burning away the forest, which reclaims the land in two years, making it necessary to clear new fields. Since it is easier to burn down the forest before the trees have properly developed, this mode of agriculture eventually devastates the forest and leads to an advance of the savanna. Another example is the eutrophication of a lake when the water's potential for self-purification is destroyed. Human interference not only harms the environment, but also affects the necessary preconditions for regenerative processes in nature.

3. *Population pressures are constantly increasing.* Two hundred thousand people are added to the world's population daily. More than three hundred years ago (1650) the earth's population totaled half a billion people. By 1970 it had reached 3.6 billion. And whereas the growth rate in 1650 was only 0.3 percent so that it took 250 years to double the

population, the growth rate now lies at 2.1 percent and the total can double within 33 years.[4]

In the hundred years between 1650 (545 million) and 1750 (728 million), the world population increased by 183 million; in the next fifty years already 178 million more people populated the earth; from 1800 to 1850 the number rose by 265 million; 1850–1900 by 437 million; and from 1900 to 1950 by 875 million. It is expected that the world population in the year 2000 will total 6.6 billion, and 9.5 billion by the middle of the next century.

Population density will also increase significantly. Given that a third of the land on the earth is cultivated, then at present 70 people are living on a square kilometer of land. In the year 2000 the population density will increase to 140 per square kilometer, and by the year 2040 to 200.[5]

4. *The return of wastes to the environment affects the quality of environmental services.* Under the law of the conservation of matter, approximately the same amount of materials taken from the environment is returned to it. The disposal of wastes in the environment degrades this public consumer good, and, along with it, the living conditions in the spaceship earth. Resources can also be affected by pollution, insofar as the regenerative functions of nature are severely compromised.

5. *The effect of wastes on the quality of the environment depends on the assimilative capacity of the medium.* Many wastes can be decomposed within the various environmental media. Bacteria reduce organic wastes if these wastes do not exceed the self-purification capacity of the body of water. The root system of bullrushes also purifies water and kills pathogenic microorganisms, raises the oxygen content of the water, decomposes organic materials, and reduces phosphates and nitrates.

6. *Nondegradable pollutants accumulate over time.* Many wastes decompose only very slowly, and others cannot be decomposed at all. To these persistent wastes belong all the toxic metals and many pesticides, such as DDT. The danger of their accumulation is that they have barely perceptible local effects when present in minute quantities, but grow to significant concentrations over time in one environmental medium or in a subunit of a biological system.

The accumulative capacity of a pollutant, then, indicates the danger of even the most incremental increase of that pollutant. A problem involving accumulation requires a solution concentrating not on the amount of the pollutant *discharged* within a certain period, but rather on the amount already present in the system. However, this approach is totally disregarded in many economic and political measures.

7. *Decisions about uses of the environment are not always reversible.* At any given time, society has a variety of options to choose from. If use A of the environment is chosen, then use B is excluded. This asymmetry of options, as it is called by Krutilla (discussed in Chapter IV), means that the use of environmental goods is fixed for a certain period of time. A subsequent change in the use of a good is sometimes techologically impossible, or institutionally very difficult to implement.

8. *Investment and technological progress can increase the availability of raw materials and can improve the regenerative function of renewable resources.* Investments facilitate the exploitation of known resources which in the past had been technically or economically beyond reach. They allow for improvements in the regenerative functions of food sources, as with marine cultivation. Technological progress can lead to the discovery of new raw materials, supply the substitute inputs which replace scarce resources in the production process, lead to a rise in the output of a given amount of raw materials, change regenerative functions of raw materials, improve the assimilative capacity of the environment, and reduce the amounts of pollutants discharged.

9. *Recycling not only decreases the wastes delivered to the environment, but also decreases the amounts of raw materials required.*[6]

THE END OF THE SPACESHIP?

How do the above-mentioned propositions interact? What are the results of their interdependence? Diagram 6 presents a simplified version of the manifold interrelationships between the economic and the environmental systems in a growing economy. Each arrow symbolizes a relation between the variables.

1) The amount of food available determines the maximum level of the population. 2) The population limit also depends on the amounts of nonfood goods produced. 3) Since a fall in environmental quality influences life expectancy, the population level also depends on the environmental quality. 4) The amount of nonfood products is influenced by the stock of resources. 5) The supply of food is also determined by the resources available. 6) Since nonfood products also include capital, the amount of these goods also determines the available food supply. 7) Technical know-how affects the amount and quality of a)resources, b)food, c)nonfood products. 8) Environmental quality is determined by the size of the population and by the production of 9) nonfood products

Diagram 6. Interrelationships between economic and environmental systems in a dynamic economy

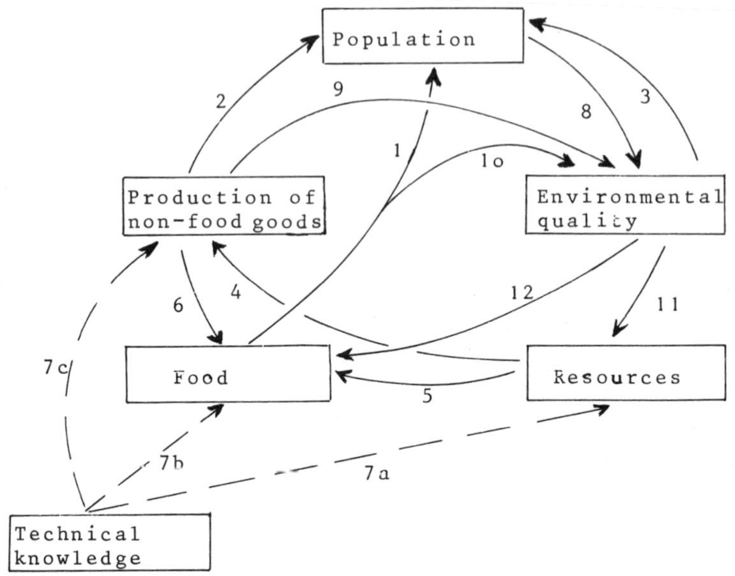

and 10) food. It determines 11) the quality and amount of resources and 12) food.

With the help of a significantly more complex system comprising over a hundred equations processed by a computer, researchers at the Massachussetts Institute of Technology[7] attempted to answer the question about the future of spaceship earth. The model includes the following variables and their interrelations: nonrenewable resources, food, population, capital as a measure of economic growth, and environmental pollution.

The model is applied to a number of possibilities based on different assumptions about the five variables. Assuming that raw materials are limited, then the system collapses when the supply of capital reaches a level which requires an enormous input of raw materials. With decreasing resources, more and more capital is used to discover raw materials. But as a result of the limited amount of raw materials, industrial growth is halted. Since services and agriculture are bound to this industrial base, they also decline. The population must drastically decrease through a rise in the mortality rate.

Assume that an astonishing technological development permits the reduction of pollutants by a quarter. In this case environmental pollution is not the critical factor in the future of spaceship earth. Now the food supply will become the major problem. The food available per capita falls as the population increases, and industrial production also falls because more capital must be used to raise food production. The death rate rises and population growth decreases.

The system also fails from pollution when, in addition to the unlimited resources and the spectacular technological breakthrough in the disposal of pollutants, a doubling in agricultural productivity occurs. Yes, even voluntary birth control will only postpone the food problem for another ten or twenty years. Finally, even if 75 percent of the resources are recycled, environmental pollution is reduced by a quarter, agricultural productivity is doubled, and if effective birth control methods are made available to the population, the system will break down. In all these cases growth will end in a hundred years and the end will come with an abrupt reduction in population through an increase in the death rate resulting from food scarcity or environmental pollution.

The MIT study concludes that survival in spaceship earth without an abrupt decrease in the population is possible in the long run only if both capital and popuation levels are stabilized. A stationary condition, already described by the classical English philosopher and economist John Stuart Mill, by Malthus with his prognosis about progressive population growth and only proportional increase in the food supply, and by Alvin Hansen's stagnation theory relating Malthus to Keynes, will become the new orientation point of economic and social policy. This stationary condition is characterized by zero growth. The population stagnates, since death and birth rates balance each other out. And the economy stagnates, since capital depletion is equal to investment—no net investment is undertaken. The given constant national product per capita is in harmony with a tolerable level of environmental pollution. The authors of the MIT study are proposing the same regimen as that propounded by ecologists: a halt to economic growth. Not an abrupt one, to be sure, but a goal to be achieved over two or three generations, an objective toward which policies should now start aiming.

THE PLANNING APPROACH

The MIT model points out a serious problem: how should this Robinson Crusoe stranded in the universe distribute the consumption of his food over the individual periods of his odyssey? Enclosed in his spaceship he must take into consideration not only the limited amount of food he has available, but also the effects that his consumption and production activities will have on his environment. More generally, how should today's known and available resources be distributed over present and future consumption? Should they be distributed evenly over future periods? Should we reserve the most important raw materials for future generations? Or should we just eat, drink, and be merry?

How can an acceptable level of environmental pollution be defined? What level of environmental quality should we seek to achieve by the year 1980?—by 2000? Should we already start to worry about the effects which are not expected until the year 2525? And should as much attention be devoted to these distant effects as to those we are confronting here and now? Will future generations not be much richer than we are today, due to economic growth, capital accumulation, and technological progress? Will they not therefore be better equipped to deal with the problems of pollution and resource depletion than we are at present? The questions about the course of the spaceship and the living conditions on board are central to the planning models being designed by economists. The variables indicated in Diagram 6 and countless other factors should be integrated into these models.

Planning and decision-making models contain:

—a *goal function* which must precisely define the social, political, and economic goals of society;

—many *constraints* which cannot be ignored. Some of these arise from the nature of the problem, i.e., they are technical or economic in character, and cannot be regulated by a policy decision. For example, the resources used in a certain period cannot exceed the amount available. Other restrictions are of a politico-economic nature and have a normative character (similar to the goal function);

—a *system of relationships* which reflect reality, including the interrelationships between the variables and the constraints.

Planning models determine the input of the resources which permit the maximization of a given goal function. In our case the resource is the scarce good called the environment. The maximum of this goal function can be examined for a single period. A period model of this type only presents the best use of the scarce resource for a given planning period. The second, third, and subsequent periods are not taken into consideration. A new optimal input must be determined for each different period. It remains unclear whether the best combination of production functions for the economy as a whole in the first period should be maintained, or whether it should be changed.

For environmental purposes, intertemporal optimization models should be developed which take several periods into consideration. In these models, too, the input of factors and the environment should be favorable on an overall economic scale. In other words, the goal function must achieve a maximal value. This maximum of the societal goal function can be calculated in two basic ways. On the one hand, it is possible to define the maximum for the end of a period. In this case the goal function in the interim period is not emphasized; at best it is included in the constraints as a minimal level to be attained. On the other hand, the maximization of the societal function can be determined for a time span covering many periods.

Both approaches lead necessarily to very different results. The maximization of the goal function for the end of the planning period is oriented toward the benefits of future generations, possibly involving sacrifices in initial periods of the plan. The maximization of the goal function over a series of periods, on the other hand, considers the value of the goal function over time. The decision about the planning horizon and minimum level of environmental quality for the planning period is a value judgment. Similarly, the weight accorded to the welfare of each generation in the welfare function is a normative decision. In the next chapter this question will be discussed again with respect to the discount rate or values over time.

The key question which planning models try to answer is: which of the possible competing uses for resources and the environment maximize the societal goal function? Opportunity costs must be examined explicitly. If the resources are allocated to sector A, they can no longer be used in sector B. If the environment is used to dispose of pollutants, it cannot be used for consumption purposes. The application of a planning model to several periods indicates that opportunity costs can also arise through the use of a nonrenewable resource in one period so that it cannot be used in future periods. Planning models for several periods must therefore determine the use of resources and of the environment over many periods in order to achieve maximum societal welfare.

The goal function which defines societal welfare is based on value judgments and subjective, not scientific, reasoning. Every member of society can assign a value to a certain level of environmental quality. A prerequisite for politico-economic decisions is the existence of a social welfare function which classifies alternative social desires. Such a function is sometimes defined independently from the values of the individual members of society. The hierarchy of societal values is then determined by a certain group, a government, a "superman," or a dictator. The authors, however, prefer the idea of integrating the values of individual members of society in the social welfare function. Decisive in such a function is the method by which these individual values are introduced into the overall hierarchy. Such methods include, for example, elections, referendums, and the assessment of private goods through market demand. Each of these methods is examined in Chapter VII.

Individual value judgments do not necessarily coincide. On the contrary, they often vary significantly, and can even be diametrically opposed. The value accorded to clean water by a scuba diver is very different from that expressed by a motorboat owner. There exists no generally accepted hierarchy of values. For this reason, the problem of the influence exerted by individual values on the welfare function as a whole remains largely unresolved.

Leaving aside for a moment the question of determination and agggregation of individual values, let us examine the factors which influence societal well-being. Which elements should be taken into account in a societal welfare function?

There is no doubt that the goods produced by an economy must be included in the measure of welfare, as consumption or per capita income, for example. The environment must also be taken into account. Some mechanism must be found to define the utility extracted by the individual member of society from the consumption of the different environmental inputs, such as the oxygen in the air and the toxic wastes in water.

Since such utility functions are difficult to identify empirically, the question can be approached from the opposite angle: what are the damages? The lower limit of the damages can be assessed by calculating the costs incurred by pollution, such as those involved in cleaning up the air and water.

If the damages can be expressed in dollars, then the goal function is the difference between the utility from per capita consumption and the damages per capita incurred by environmental degradation.[8] Other goals are determined by the minimum demands formulated by political actors for a planning horizon. This would include objectives expressed in terms of a minimal capital stock, a minimum level of environmental quality, and a minimum amount of raw materials which must be preserved for future generations. Decision makers should use more realistic welfare models which would include in the goal function the gross utility of consumption, the costs of treating and disposing of wastes, the costs of modifying the environment to avoid pollution, the damages which occur in spite of all measures of environmental protection, and the opportunity costs of production factors.

Decision-making models, then, include a goal function and normative restrictions which set minimum levels for existing resources, capital goods, and environmental quality. They also include many other normative or technical constraints and information. The complex interrelationships between the environment and the economy illustrated in Diagram 6 should be worked into this model.[9] In particular, the following should be included:

1. Production functions indicating the inputs which are required for the production of individual goods, such as food and nonfood products, including capital.
2. The stock of resources available for production in a given planning period.
3. Information about the conditions necessary for the regenerative functions of the raw materials and the extent to which a given category of raw materials can be expected to regenerate itself.
4. Functions which provide information about the amounts of pollutants generated at particular levels of production and consumption.
5. The change in the production matrix resulting from technical progress, investment, and the recycling of material.
6. Costs of transforming one type of waste into another, and the costs of alternative disposal processes, such as incineration and waste composting.
7. Models which describe the diffusion and transformation of wastes in the various environmental media.
8. Damage functions which permit the evaluation of the effects of pollutants on the ecological system.
9. Hypotheses about the determinants of the population level and about the development of the demand for food, other private goods, and public goods, including environmental services.
10. Emission standards for particular activities and ambient quality standards for the environmental media in each period.

DOOMSDAY AND ADAPTIVE MECHANISMS

The planning approach is a heuristic device to analyze the problem of long-term environmental policy. The advantage of this approach is that it draws together all components of the problem. However, this provides no information as to whether and how a plan can be implemented. Who should plan and who should determine the goal function? Should environmental planning be conducted at the national or international level? What lessons are to be learned from planning experiences in Eastern Europe, the European Communities, or Hitler's National Socialism?

Before embracing the planning approach, it is worth investigating the extent to which endogenous adaptive mechanisms can help resolve enviromental problems and provide reprieve from the doomsday predictions discussed above. Does the MIT planning model take these mechanisms into account? If not, might the results have been less pessimistic had these processes been included in the calculations?

—The MIT model accords too little weight to substitution processes. The economic system can adapt to scarcity by substituting factors of production and consumption. The process is directed by prices. This built-in adaptive mechanism in the market system makes active government planning unnecessary in many cases.

—Technological progress is defined in the model as a unique improvement in the use of resources and the reduction of environmental pollution. But this assumption does not take into account alternative premises about technical progress. Technical progress encompasses the development of new production processes, new goods, organizational improvements, the discovery of new resources, the perfection of exploration and extraction technology, and the recycling of material. For example, if the extraction of minerals from the sea bed is technologically possible and remunerative in view of today's prices for raw materials, we are talking about an increase of 30 percent in the resource base. In the course of constructing the Alaska pipeline technicians developed new processes and equipment for extraction which could be used in the Antarctic. A major increase in the prices of raw materials and in emission charges will create incentives for the development of new technologies for emission abatement, recycling, and discovery of new raw materials. The prospect of profits also motivates technological development. It cannot be overlooked, however, that such developments can entail new dangers for the environment if possible negative effects are not considered in advance.

—Both the producer and the consumer take advantage of substitution options and incentives to adapt to scarcity by using different goods or revising their "consumption technology" (e.g., heating technology for homes).

—A decisive assumption of the MIT model is that no major change will occur in the social system. The model proceeds from the observable conditions of the past and does not take into consideration, for example, future feedback and modification of political decision making. Imagine that the development described by the MIT model were about to take place. In such a case, organizational changes, such as a restructuring of the goods produced by shifting the burden of environmental costs on the polluter, would have immense effects on environmental quality. The dichotomy between the need for government intervention and the institutional problems of government planning is examined in greater detail in the following chapters.

The planning models discussed here are constructed on approaches to programming theory[10] and control theory.[11] Chapter VII concentrates explicitly on the role of the environmental problem in the economic system and evaluates cost-benefit analyses as a tool for pragmatic decision making.

NOTES

1. K. Boulding, *Economics of Pollution,* New York, 1971.
2. R. d'Arge, "Essay on Economic Growth and Environmental Quality," *Swedish Journal of Economics,* 73: 25–41 (1971).
3. D. Meadows et al., *The Limits to Growth,* London, 1972, p. 46.
4. Ibid., p. 26.
5. W. Fuchs, *Formeln zur Macht, Stuttgart,* 1965, p. 71.
6. D. W. Pearce and I. Walter, "Resource Conservation," *Social and Economic Dimensions of Recycling,* New York, 1977.
7. Meadows et al., op. cit., p. 45 et seq.
8. If it is impossible to assess environmental quality, then this factor must be treated in planning models as a normative restriction which cannot be ignored.
9. C. S. Russell and W. O. Spofford, Jr., "A Quantitative Framework for Residual Management Decisions," in A. V. Kneese and B. T. Bower, *Environmental Quality Analysis, Theory and Method in Social Science,* Baltimore, 1972, pp. 115–118.
10. W. O. Spofford, C. S. Russell, and R. A. Kelley, "Operational Problems in Large Scale Residuals Management Models," paper prepared for the conference on "The Economics of the Environment" organized by Resources for the Future and the National Bureau of Economic Research, Chicago, Nov. 10–11, 1972.
11. R. C. d'Arge and K. C. Kogiku, "Economic Growth and the Environment," *Review of Economic Studies* 40:61–77(1973).

Chapter VII

A Cleaner Environment: Costs and Benefits

We must come to understand that protecting and maintaining the environment is not a cost, but an investment.

Maurice F. Strong
Catalyst, 3/76

Purists demand an environment in which the air no longer contains sulfides and in which the Ohio River flows clear as a mountain stream, an environment in which there is once again space for wild animals and where the rarest flowers bloom. It cannot be denied that such a paradise is desirable, but can it be achieved?

Such a world, the realists would answer, is only possible if society is prepared to accept the costs: renounce possible increases in goods produced, effect radical changes in consumer behavior, sacrifice jobs, and accept lower tax revenues for financing public goods. What must we sacrifice if we want a cleaner environment? How much growth are we willing to forfeit to improve environmental quality? What is the tolerable level of pollution?

THE COST-BENEFIT CRITERION

When faced with such questions, economists have recourse to a method of decision making designed initially to deal with large public investment projects, particularly dams, and which has now become a tool in political economics—the cost-benefit analysis. Imagine that the government must decide which dams to construct with a given amount of funds. Since there is not enough money to finance all projects, one dam must be chosen from among all those proposed.

A dam has many positive, desirable effects. It permits the production of electricity, it allows the regulation of a river's water volume, and contributes to flood control; it makes rivers navigable, helps irrigate agricultural land, replenishes the ground water supply, and can contribute to the drinking water supply. A dam also entails costs, like initial investment and constant maintenance costs. A comparison of the costs and benefits of various projects permits the identification of the most profitable one, the one in which the difference between the costs and the benefits (i.e., the net benefit) is at a maximum.

The systematic comparison of the costs and benefits can also serve as a basis for decision making with respect to environmental conservation measures. In this area, cost-benefit analysis represents a pragmatic method for solving the problem of competing uses of the environmental services. It is a tool for choosing the "right" use of the environment. Cost-benefit analyses regard the use of resources for the improvement of the environment as a cost because these resources cannot be used for the production of other goods. Costs are, therefore, foregone production, i.e., a decrease in the amount of goods produced. In a growing economy this represents a reduction of the real rate of growth vis-à-vis the potential rate of growth. The benefits of an environmental protection measure stem from improvements in the quality of the environment, such as cleaner drinking water and the removal of pollutants from the air.

The cost-benefit criterion can also be applied to the question posed in Chapter V about "growth versus environmental quality," assuming that economic growth is desired. As long as the increase in the goods produced is valued more highly than the damages caused to the environment in the process, then economic growth will be pursued. However, at the point where environmental deterioration is assessed to be greater than the value of an increase in production, economic growth must be halted or restructured. By assigning to a given environmental medium the use which minimizes the overall societal damage, the cost-benefit analysis can determine the particular balance of economic growth and environmental quality which reflects societal priorities.

Unlike the allocative decision making of private enterprises, the cost-benefit analysis attempts to assess the benfits of a given or improved environmental quality. It does not assume zero price for environmental services. Yet, is not the attempt to evaluate the benefits of the environment an impossible venture? How can environmental goods like air and water be assigned a value? Trying to hang a price tag on such goods only seems to verify Bertolt Brecht's criticism of the capitalist system: "The rise of this monstrosity is inexorable, converting nature into merchandise; even air is marketable."

The fact remains, however, that every society, including a socialist one, faces the problem of very different, competitive uses for the environment. A lake can provide swimming facilities or drinking water. The water from a river can support either industrial production or human consumption. A cost-benefit analysis must be conducted to evaluate the competitive uses of the environmental services and to identify the most desirable of the activities. Cost-benefit analysis can thus be interpreted as a pragmatic approach to judging the net benefits of a policy measure. Let us examine this decision-making procedure more closely.

THE COST-BENEFIT CRITERION AND ENVIRONMENTAL PROTECTION

If damages avoided are considered a benefit from an environmental policy measure, then the assessment of benefits must be based on the damage function. Diagram 7 illustrates a very simple case in which only one waste and one pollutant are depicted.

The amount of wastes produced by one or more firms in a region is plotted on the right axis. These wastes must be assessed. The diffusion function (a waste's decomposition and processing by the environment and its transportation to other media) is represented in the lower right quadrant of the diagram. The lower axis indicates the amount of pollutants, and it can be interpreted as decreasing environmental quality. The transformation function thus represents the level of environmental quality resulting from a given amount of waste.

Environmental quality is assessed in the lower left quadrant in that certain levels of quality or certain amounts of pollutants are assigned damages expressed in terms of dollars. These damage values can then be coordinated with the amount of waste indicated in the upper right quadrant. The coordination is achieved by a 45° line in the upper left quadrant. The amount A_1 of wastes produces quantity P_1 of pollutants. This causes damage D_1 so that the coordinate point of waste and damage is provided by point A' in the upper right quadrant. Other points can be

Diagram 7. Determining the damage function

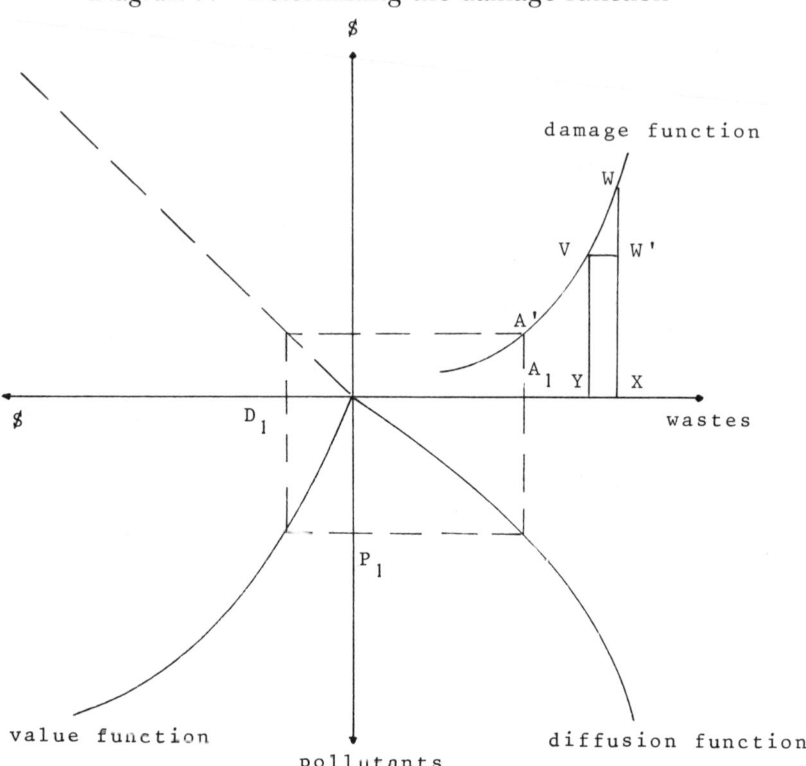

obtained in an analogous manner. The curve in the upper right quadrant then gives the damage function in dollars with reference to the amount of waste.

The damage function in the upper right quadrant permits the assessment of an improved environment. It indicates how much the assessed damage decreases when the amount of wastes emitted is reduced. If, for example, the wastes are reduced by the distance YX, then the damage decreases to the extent of WW′.

The reduction of wastes also creates costs, as illustrated in Diagram 8. The function (read from right to left) indicates that the more wastes are reduced, the more expensive it becomes to dispose of each additional unit of waste. The cost of reducing wastes from 100 percent to 99 percent is less than for a reduction from 30 percent to 29 percent (point C′). Costs rise steeply if all wastes must be removed. The acceptable level of pollution has been reached when the net societal benefit is at a maximum, i.e., when the difference between gross utility and the costs of environmental measures are at a maximum, thus B − C = max! Usually, the benefit is measured by the damages avoided. If D_0 represents the damages before environmental measures are implemented, and D_1 those afterwards, then the decision rule is $D_0 - D_1 - C$ = max! Since the initial damages are independent of the environmental measures, they represent a constant factor in the equation. The maximization can then be resolved by minimizing the damages and the costs of abatement. The equivalent decision rule $D_1 + C$ = min! therefore requires the minimization of total societal damages. This minimum is represented by point R in Diagram 8.

The equivalent decision rules presented here also lead to equivalent marginal conditions. The maximization of the net benefits represented by B − C requires that the marginal benefit of an environmental measure equal its marginal costs. If the benefit is measured by the damages avoided, then the marginal costs of the environmental policy must equal the avoided marginal damages, as in the case of minimizing the overall societal damages. Measures of environmental protection are worthwhile as long as the overall societal damages decrease, and the damages de-

Diagram 8. Estimating the costs of pollution abatement

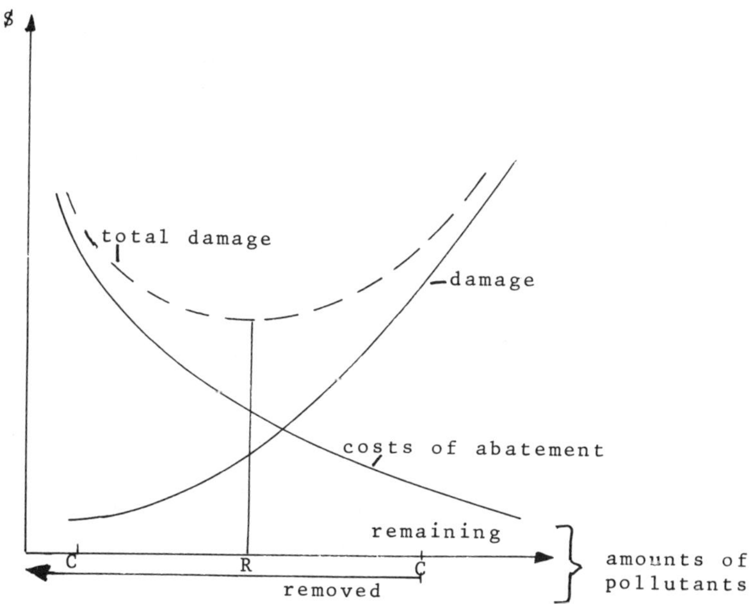

crease as long as the marginal benefit of pollution abatement is greater than the marginal costs of abatement.

As indicated throughout, this decision-making process only compares benefits and costs; it does not derive them. Since cost-benefit analysis in the area of the environment functions only when society has first assessed the value of the activities in question, it is necessary to further explore how these values can be discovered.

HOW MUCH IS THE ENVIRONMENT WORTH?

The simplest possible form of assessment is to assign infinite value to one use of the environment and zero value to all others. For example, pure air will be assigned infinite value, and all other possible uses of air will receive zero value. If this value judgment is put into practice, then the environment can only be used for life-sustaining activities. Humans would then have the consolation of a clean environment—even if they had to starve for it.

A less stringent formula for assessment would be the establishment of a hierarchy of values for the possible uses of the environment. The provision of drinking water, for example, would be valued more highly than the use of a lake for swimming. Such a scale would be an important and, theoretically, sufficient basis for decision-making.

In fact, pragmatic decision-making by cost-benefit analysis would not be necessary if we could assume that there existed a social welfare function that comprised all the relevant economic targets. Let W denote societal welfare, and let Y denote national income, P price stability, E employment, Z balance of payments equilibrium, and α other policy targets (e.g., income distribution). The welfare function would be $W = W(Y, P, E, A, \alpha)$.

Such a formula is theoretically very elegant, but in reality it is hardly practicable. Who, for example, can unilaterally assign infinite value to pure air? Who will put that value judgment into practice? Or who will establish a hierarchy of values for all of society upon which all wide-ranging decisions should be based? Clearly the welfare function must be developed from the preferences of the individuals in society, not from the opinion of a single individual or group. But we do not know how to specify such a function. No such function indicating "more important than" values has been established and accepted for all members of society. Therefore, we must look for methods which permit an assessment of the benefits of environmental protection measures.

1. Market prices

The usual way of assessing the value of goods is to use their market price. Market prices indicate how much individuals are willing to pay for a good, or which other possible uses for their income people are willing to renounce in buying a particular good. If the market price becomes the basis for the evaluation of an individual's priorities, demand determines the value of a good.

But since the environment is a public good, market prices by which to evaluate it rarely exist. How, then, should the value of a particular level of air quality be measured? How can the value of its improvement be identified? How does the value placed on the air alter when air pollution increases enough to make the smog report a standard part of weather forecasts?

Market prices can be used only to a limited extent in those cases relating to the assessment of production and property damages. If the technical dimensions of the damage function are known—and this information is a precondition for any assessment—then production and property damages can easily be expressed in terms of dollars.

2. The interview approach

It is imaginable that each individual could express "more important than" functions for a list of uses of the environment, and these lists could then be synthesized for society as a whole. This raises two major problems. First, it is not clear that interview techniques adequately reflect opinions and values; there have been sufficient examples of grossly false prognoses on voting behavior to cast serious doubts on their reliability.[1] Secondly, even if the individual hierarchies can be identified, the problem still remains of obtaining an overall hierarchy for society. Aggregation of individual values into a single hierarchy is of questionable practicability.

3. Willingness-to-pay approach

This alternative seeks to determine how much an individual is prepared to pay for a cleaner environment.[2] The analysis of willingness to pay attempts to establish the preferences of individuals through interviews or other mechanisms such as voting or referendums by posing such questions as "How much would you be willing to pay in order to reduce air pollution in your town by 10 percent?" Or, "How much would you be

willing to pay to reduce air pollution to medically established levels of safety?" The question can just as easily refer to the reduction of specific pollutants in water or to the maintenance of a given level of environmental quality.

The analysis usually requires more than a single relation between environmental quality and the willingness to pay. A demand function must be established by assigning alternative environmental qualities to alternative degrees of willingness to pay, as is illustrated in Diagram 9. Presumably, the willingness of an individual to pay decreases with the level of pollution. Individuals would be willing to pay more for the reduction of pollution from 100 percent to 90 percent than for the reduction from 30 percent to 20 percent. The shaded portion of the diagram relates the pollutants removed to the net benefit drawn from a reduction of pollution by 25 percent. The overall benefit to society is obtained by adding together in terms of dollars the benefits experienced by all the individuals.

An individual's willingness to pay is determined by a series of factors. In addition to the person's attitude toward society, which is also expressed in different attitudes toward taxation, the information the individual has about the effects of environmental pollution plays an important role. The question about willingness to pay for the maintenance of a desirable situation will yield a different assessment than a question of willingness to pay for reestablishing a previous level of environmental

Diagram 9. Assessment of the willingness to pay for pollution abatement

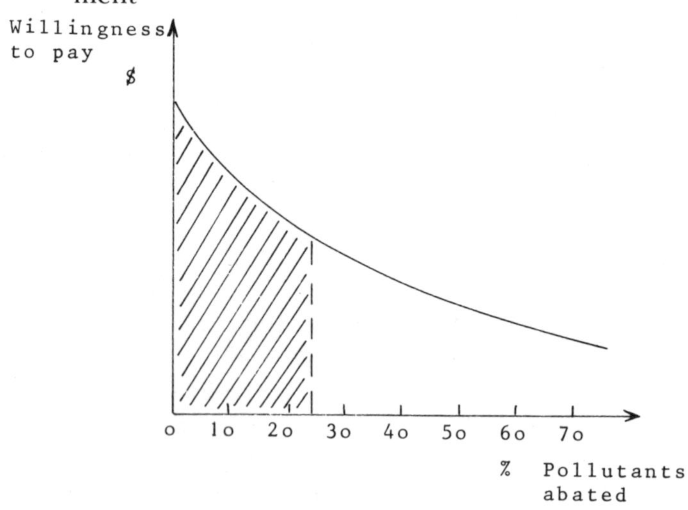

quality. The information level of the individual in these two cases differs. Those people living in a polluted area have experienced the damages and will assess the consequences of pollution differently than people living in a clean area, unless, of course, they have adjusted perfectly to the pollution. Further, the person must be familiar with the damage functions described in Chapter III. Individuals must know about the damages caused by certain levels of pollution and how health, environmental services, production, and capital goods are affected by environmental deterioration.

The way in which pollution abatement measures are financed also influences a person's willingness to pay. An individual will never overlook the problem of the free rider, for example. If people known that the free rider can be forced to contribute to financing envirnomental protection, they might raise the amount they are willing to pay for a cleaner environment.

How much an individual is willing to pay for environmental protection is a hypothetical question. It is very difficult to obtain valid answers. As indicated above, the problem of the free rider affects the individual's assessment. A second source of distortion is fear that the survey might later be used as a basis for establishing a tax. In this case, the individual would provide a low estimate, in order to avoid paying high taxes later. Thirdly, a person might express a high value in order to lend greater weight to a particular use of the environment, and thereby bias the results of the survey. If it is not possible to establish the real value of the individual's statement, then the relevance of the analysis is questionable.

In order to obtain valid answers, it has been suggested that the individual should not be given information as to the type of financing measures to be undertaken, so as to give the impression that the free-rider problem has been resolved.[3] The person interviewed does not know whether the improvement in environmental quality will be financed through general taxes or through particular fees to be paid by the individual users. Such uncertainty, according to this thesis, will force the interviewees to express their true values. Whether this is the case remains an open question.

Willingness to pay varies with the type of public good involved. Many goods are spatially limited. Air pollution, for example, can be limited to a particular region. The more spatially restricted a public good, the easier it is to distinguish it from other public goods, the more definable the group involved, and the greater the willingness to pay. It will be fairly easy to remove a dump on the outskirts of a small community by dividing the costs among the small population. However, it is to be expected that the willingness to pay for cleaning up the atmosphere will be rather low. If this thesis is correct, then global public goods which cannot easily be spatially defined will be undervalued.

Willingness to pay depends on the frequency and intensity of use. Individuals who do not demand a particular environmental good will show very little willingness to pay for its protection. Who will be prepared to pay much for the purification of sewage knowing that he will not be using the water again, but that it will be used downstream? And who will be willing to contribute to developing a park he cannot get to?

In addition to the spatial delimitation of the public good, the intensity of need also plays a role in willingness to pay. One can imagine that certain groups accord greater importance to a given level of environmental quality than do others. Individuals with heart and lung ailments, bronchitis, asthma, and other breathing difficulties will value clean air more than will healthy individuals.

Lastly, willingness to pay varies with income and wealth. This is not because wealthier people can afford to pay more, but because individuals with a higher income can enjoy a broader range of consumer goods and will thus probably value a cleaner environment more than those individuals in a lower income bracket. We will examine this aspect of environmental protection further in Chapter VIII.

The reader might have the impression that the authors want to make Brecht's warning a reality by placing air and water on the open market in order to resolve the environmental conflict. This is not our aim. We are only proposing a method of assessment, not an approach to a final solution. The evaluation of environmental services by households through willingness-to-pay analyses is only designed to calculate the benefits of the reduction in pollution. The fact that these analyses are fraught with problems must be clear by now.

4. Cost assessment

Another approach to the assessment of the environment is based on the assumption that environmental quality should be assessed only according to the costs a person is willing to absorb in order to achieve it.[4] Here it is necessary to distinguish between transportation costs, rents or property prices, and the costs of pollution abatement. Accordingly, the value of environmental quality is expressed in migration patterns. Generally, the income levels in agglomerated, polluted, industrial regions are higher than in less environmentally degraded areas. Individuals who leave a polluted area "vote" against it. If environmental quality as a determinant of regional mobility can be sufficiently isolated from other factors, such as differences in pay, social attachments, etc., then an approach has been found for the assessment of environmental quality. The foregone income and the costs of the move—including everything from the movers' fees to the time spent looking for a new house—are indicators of the costs an individual is prepared to absorb for a better environment. If data from various regions are available, then different levels of environmental quality can be correlated to different foregone incomes, thereby producing an assessment function for environmental services. A study of mobility behavior, however, can only provide information about those aspects of environmental quality which are spatially limited. No one will change jobs because of increased levels of sulfur dioxide in the atmosphere as a whole. So this approach, too, will underestimate the value of global environmental goods.

A cost evaluation can also be conducted for public services such as natural parks by ascertaining the travel and related costs involved in consuming this good. If, for example, it must be decided whether a natural park should be closed and mined for newly discovered coal or copper, one possible criterion to establish the lower limit of value could be the number of people who would have to drive the additional miles to get to an equivalent park. It must be noted, however, that this approach cannot be used to measure the value accorded to a landscape which is gradually destroyed, for short-term changes caused by the expansion of suburbs, new roads, and slowly increasing air and water pollution go almost unnoticed. In this case it would be necessary to ask the older generation, who still remember the initial condition, what benefit has been lost through the change in the landscape.

The prices of housing and land can also be interpreted as an indicator of environmental quality. It is to be expected that areas with better environmental quality will be preferred by buyers. This preference increases the demand for property in such areas and thus raises the land prices.

Ridker[5] has attempted to relate the value of land and buildings to air pollution. The concentration of sulfur trioxide in the air was measured in various parts of St. Louis. Several other factors were taken into consideration, such as the density of the population, proportion of new buildings, and driving time. The result was that a reduction of the level of sulfur in the air to a specified lower level would increase the value of houses in St. Louis by about $85 million. At an interest rate of 8 percent this means a $7 million increase in capital income per year.

As in the case of migration, the analysis, of property values also depends on the sufficient isolation of environmental quality as a factor influencing the value of land and buildings. This means that the influence of such factors as type and average age of the residential area and social status must be isolated. These studies involve significant difficulties. First, since buildings and land are not sold very often, market prices are seldom available. It is then necessary to rely on approximations derived from tax values. Secondly, this approach assumes that prices really do reflect the values attached by the buyers to environmental pollution, that the buyers consider pollution to be a significant criterion in the decision to buy.

The assessment of environmental quality through the prices of residential property is similar to the approach based on the costs of moving. Both measure the quality of the environment indirectly by measuring another element. It is assumed in both cases that the property prices reflect the behavior of the individual with regard to a change in environmental quality, and further, that this behavior provides an adequate basis for assessment.

A cost assessment of environmental quality can also be expressed in terms of the costs of pollution abatement or the reparation of a destroyed environmental good. For example, the costs of purifying polluted water to provide drinking water can be considered a minimum value level for drinking water. The maintenance costs for buildings, such as the increased need for cleaning façades as a result of air pollution is also a means of assessment. Another is the costs of replacing steel parts in buildings because of the corrosion caused by air pollution.

Computing costs incurred by environmental degradation as a lower limit of value for environmental services has also been suggested for health damages when the preferences of the individual are not identifiable. Assume that the damage function presents only an association between pollutants and sick days or sickness probabilities. The social costs of illness, which include doctors' fees, costs of treatment, hospital bills, and lost income represent in this case a lower limit for the assessment of these effects. Similar studies are conducted on deaths related to environmental degradation. Here the income lost by the individual who has died prematurely is considered a "cost factor." In the United States the Council for Environmental Quality estimates from data in 1963 that a 50 percent reduction in air pollution in major cities would reduce health care costs by $2.08 billion a year.[6] If the proposed standards of air quality can be achieved, $6 billion would be saved in health care costs yearly.

It is certainly distasteful to express death and sickness in terms of dollars. There are two views on the matter. One holds that the consideration of health care costs and death as a lower limit results in an improvement of environmental quality that is profitable from a purely economic point of view; pragmatism in terms of dollars is therefore more effective than philosophical lamentation. The other view is skeptical. If one is able to assess illness and death only with a very doubtful lower limit, then such damage assessments should be avoided altogether. The damages should be expressed in nonmonetary terms as cases of sickness and death and thus be kept separate from other calculations of benefit and costs. Additionally, a statistic intended to represent a lower limit of costs may be misinterpreted. Politicians accustomed to slogans and indices are apt to see in this number the correct estimate, an approach which would systematically undervalue the utility of environmental protection measures.

The reader must be warned about a logical trap. The cost assessment of environmental quality—when it involves the cost of pollution abatement—uses the costs of abatement as a proxy for environmental value. The precondition for this procedure is that the desired level of environmental quality is given, for example in the form of a fixed target. This type of cost assessment, therefore, presupposes a desired level of environmental quality, and cannot be used in a cost-benefit analysis for the determination of a desirable level of environmental quality.

While the assessment of health damages and negative effects stemming from the environment poses a major problem, the computation of damages to production and property does not. Once the damage functions are known in technical dimensions, then the damages to production and property are easily derivable in terms of dollars through market prices. Say that one could have a damage function in which agricultural productivity depends on the level of phosphate extraction, a case that was examined in Florida. In this situation, the fall in agricultural productivity only needs to be multiplied by the market prices for the foregone products. Since productivity figures were not available, Crocker[7] analyzed the effects on phosphate extraction on the price of agricultural land in Polk County, Florida. He assumed that the price of agricultural land implicitly reflects the land's productivity in citrus groves, animal farming, and the influence that phosphate extraction has on these activities.

There are, then, different methods of assessment for the various types of damages. Knowledge about the damage function is indispensable for every approach. If this is known in nonmonetary terms, then property and production damages are the easiest to establish because it is sufficient to apply the market prices. The problem is much more complicated in the case of environmental consumer services. Health damages are even more difficult to deal with. We have hardly touched on ecological damages about which little is known in terms of the whole ecological system. They are very hard to assess. One possibility is to compare past and present production figures. Whereas in industries like fishing and forestry the reduction in productivity can be determined, it is impossible to estimate the costs of cleaning the atmosphere of excess carbon dioxide or of producing oxygen when forests and sea plankton have been destroyed.

The practical application of the assessment methods described above certainly entails significant problems. A mere listing of the possible approaches does not mean that the problem has been solved. But it should be clear that methods of assessment do exist and that the evaluation of environmental quality is not impossible. If none of these approaches appear satisfactory, it must be remembered that political decision makers and the ballot box arrive at sociopolitical assessments in many politico-economic decisions, such as those related to the conflicting goals of full employment and price stability.

In order to resolve the problem of environmental pollution we cannot patiently wait until the economists have checked whether a consistent, transitive societal assessment function based on individual preferences exists. If necessary, society must assess and take action through the politician.

THE INSTITUTIONAL SETTING OF ENVIRONMENTAL EVALUATION

In the preceding section we examined several methods of evaluating environmental quality. Let us now turn to the institutional aspect of the question. Which institutional arrangements can be used to establish the value of environmental quality?

1) *Elections.* The preferences of individuals can be aggregated through elections. Usually, though, too little is known about the position of a party on environmental issues and the costs of measures it might implement. In addition, voters trade off between the parts of the party program that appeal to them and those which they either do not like or

do not find important, so that the preferences accorded to individual environmental goods are not revealed in election results. Other problems in this approach are that the free-rider question remains unresolved and the consistency of an aggregated welfare function, as indicated by Arrow, is not certain.

2) *Referendums.* The most promising method of discovering the "real" preferences of individuals is a referendum of the type conducted in some states and communities in the United States and in Switzerland. The electorate votes on different projects such as road construction and school expansion. Various methods of financing the projects are submitted to the voters, such as government loans or tax increases. The entire process whereby various social groups urge the voters to vote "yes" or vote "no" must be interpreted as a political expression of popular demand, and it certainly represents an important measure of assessment. Referendums can be a basis for assessing different environmental damages, particularly when individuals must vote on how to use their money, such as by approving a plan to build a water treatment plant that will be financed through local taxes. Here, too, however, the problem of the free rider cannot be eliminated. If referendums contain no financing options they merely aggregate individual preferences and do not provide information as to whether the voters consciously report false preferences.

3) *Political decision-making processes.* In analyzing the development of environmental goals, it becomes necessary to take into account the voting behavior of political parties, interest groups, bureaucracies, and voters who act under specific institutional conditions. For instance, it can be assumed that parties maximize votes, and the implications this behavior will have for the goal of environmental quality can be analyzed. This means that we can study the value accorded to environmental quality and to which specific measures of environmental policy the vote maximization behavior of political parties will lead. Alternatively, it might be assumed that political parties maximize utility, i.e., that they attempt to achieve basic party programs over a longer span than one legislative period (while also trying to remain in power). These questions can be analyzed in the framework of the "new political economy."[8] In this context the coalitions of political groups become important and regional interests play a determining role in the choice of objectives. The regional aspect of environmental goods therefore becomes relevant. Federalism is also an element which is significant in the aggregation of individual interests.

4) *Bargaining.* Coase[9] has proposed that environmental services can be allocated through a bargaining process. Like Dales,[10] he believes that the environmental services are misallocated because the environment is treated as a public good for which no rights of use exist. Coase therefore suggests that the polluter or the victim be assigned rights of use. If the victim has the right to use the environment, the polluter might be able to persuade him to accept a lower level of environmental quality in exchange for compensation. This assumes that the damages to the environment can be translated into monetary terms. The victim will be willing to accept pollution as long as the compensation per additional unit of emissions is greater than the marginal damage. For his part, the polluter is prepared to offer compensation for use of the environment as long as the compensation is lower per unit of pollutant than the marginal costs of abatement.

On the other hand, if the polluter has the right to use the environment, then the victim must pay to abate pollution. The victim's willingness to pay depends on the marginal damage avoided. The polluter must receive compensation payments which exceed the costs of abatement, since his bargaining position is determined by the marginal cost curve of pollution abatement. The compensation payment is specified by the point at which the marginal damage curve and the marginal cost curve intersect.

The Coase theorem shows that the problem of the allocation of environmental services can be dealt with through rights to the environment (if the conditions assumed by the theorem are met). According to this theorem, the same environmental allocation will result whether the polluter or the victim are assigned the rights to the environment.

The Coase theorem has been the subject of a great deal of discussion among economists.[11] It has been criticized on several points. Although it is true that the same allocation of environmental services is achieved with both approaches, the distributional effects are different. This raises the question of efficiency versus equity, at which we will look more closely in Chapter 8. The costs of the transaction also affect the allocation of the environment. Further, only very rarely can a single polluter and a single victim be identified. The allocation of the environment is also affected by the differing bargaining positions of the parties involved. The problem of the free rider plays a role, too.

5) *The Clarke tax.* This represents one of the recent attempts to deal with the problem of the free rider. The Clarke tax[12] involves asking each

individual to declare his willingness to pay for two alternatives, A and B. Call S_A the willingness to pay of all those people who prefer A to B, and S_B that of those who choose B over A. Society chooses A if $S_A > S_B$, in other words, when the willingness to pay is greater for A than for B. In order to make certain that individuals are not falsely reporting their preferences, a Clarke tax[13] is imposed on the individuals whose vote affects the result. They pay the difference $S_A^+ - S_B^+$, whereby S_A^+ and S_B^+ are calcu- without the individual's willingness to pay. Each individual thus has the option either to leave the result as it would be without his vote or to change the result at the price of a net loss for others. A false answer does not benefit the individual in any way. If the value of a result preferred by the voter is less than the net value for others, then the voter will leave it as it would be without his vote. There is, therefore, no reason to falsify an answer. If, on the other hand, the value for the voter is greater than the net value for others, then the voter will want to influence the result, and he has no reason to hide his preference. The Clarke tax thus represents a social choice mechanism which forces the individual to reveal his true preferences.

Tideman and Tullock consider this demand-revealing process to be immune to individual maneuvers and they believe it avoids the problems involved in developing a consistent and transitive, nondictatorial aggregation of individual values. The voter is asked to do more than simply arrange the alternatives into a hierarchy, he must also assess the alternatives. However, if the voters form coalitions they can undermine this new social choice mechanism. Further, the Clarke tax creates a surplus in the budget, and it is not clear how this excess should be handled. Should it be taken out of circulation, or should it be redistributed? The latter proposal must be examined closely to avoid affecting the incentives for providing unbiased answers.

6) *The Groves-Ledyard mechanism.* This mechanism[14] is related to the Clarke tax. Individuals inform the environmental agency of the level of environmental quality they desire, but do not assess it. The environmental agency then undertakes a positive or negative compensation. If an individual desires a cleaner environment than the average, he is burdened with paying for it. This can lead to a change in his wishes. The communication process between the environmental agency and individuals is iterative. The result is that in the end all individuals report the same level of desired environmental quality. Elections, referendums, and the Clarke tax do not lead to this situation. In elections, for example,

the desired level of environmental quality is determined by 50.1 percent of the voters. However, in order to succeed in attaining this unitary level of demand, the compensation function of the Groves-Ledyard mechanism must be very strong.

THE OTHER SIDE OF THE COIN

If environmental policy measures had benefits only, we would not have to discuss the matter any further. Unfortunately, they also incur costs. Of these, the least complicated to determine are the costs of equipment for pollution abatement or reduction and the constant maintenance and service costs. Such costs appear for households (emission standards for automobiles), for firms (emission standards for production processes, definition of product quality), and also for the government (emission standards for sewage plants). Like the benefits, the level of the costs depends on the measures chosen, such as taxation, emission standards, etc.

Abatement costs are expressed in terms of opportunity costs because resources to implement environmental policy measures must be withdrawn from production. Abatement cost functions indicate that these costs increase progressively with the level of abatement.

For instance, Diagram 10 illustrates the cost abatement function for the sugar industry. The vertical axis represents the costs of the reduction of one unit of biochemical oxygen demand (BOD) as measured in pounds. The horizontal axis (read from right to left) measures the reduction of BOD in pounds and percentages. The reduction of 10,000 pounds, or 30 percent of BOD would cost less than a dollar per pount of BOD. However, the reduction of 65 percent of BOD would cost $21; 80 percent would cost $30 per pound of BOD, and a 95 percent reduction would cost $80 per pound. Similar results have been obtained in other studies and for other sectors.

Table 5 estimates the abatement costs of the United States for 1975 and 1984 (in billions of 1975 dollars). The total costs for the improvement of environmental quality in the United States were estimated at $31.7 billion in 1975, and it is expected that they will reach $69.2 billion in 1984. Expenditures are expected to be $486 billion between 1975 and 1984. The projections of the CEQ indicate that the costs of improving environmental quality in the United States will be 2.2 percent of the GNP at the end of the decade if legislation for environmental protection

Diagram 10.[15] Cost Abatement Function for one Sugar Industry

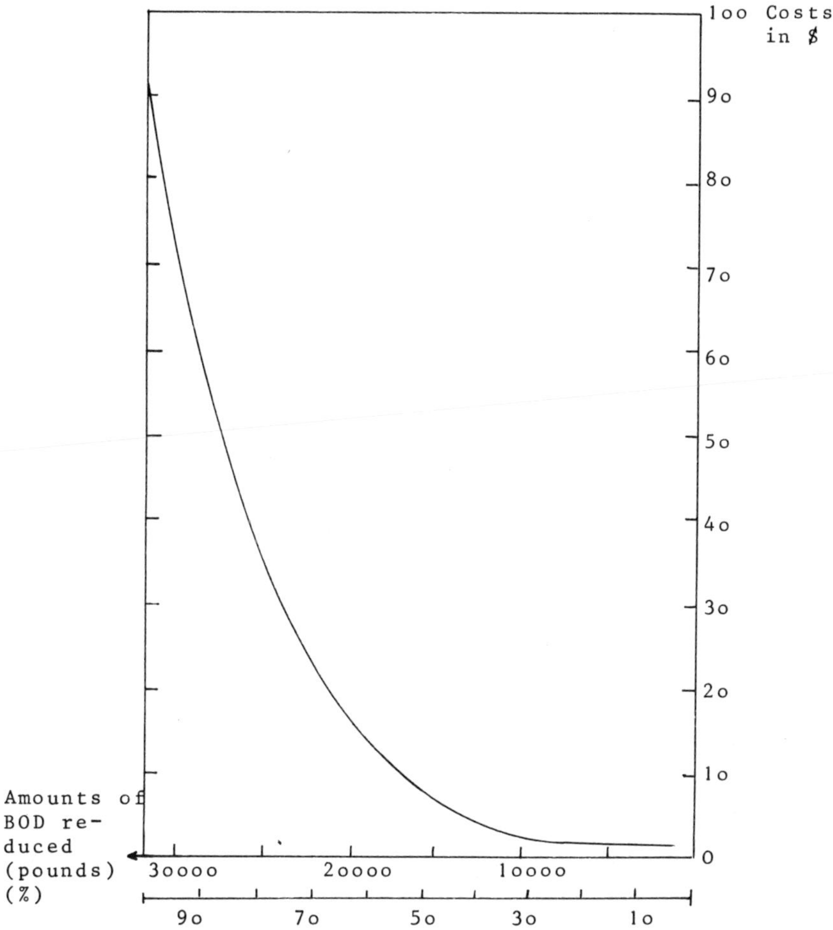

is enacted.[17] This does not include the costs of reducing noise from aircraft, land reclamation, or surface mining.

The expenditures listed in the table are only very rough estimates, but they do provide insights into which resources for environmental measures are withdrawn from production. They therefore measure the foregone GNP. In addition to these figures are costs which are documented by the reduction in the achievement of other goals. Measures of environmental protection affect 1) price stability, 2) full employment, 3) balance of payments, and 4) regional and sectoral goals.

Table 5.16 U.S. Pollution Abatement Costs for 1975 and 1984 (in billions of dollars)

Pollutant/source	1975			1984			Cumulative (1975–1984)			
	Operation and maintenance costs	Capital costs[b]	Total annual costs[c]	Operation and maintenance costs	Capital costs[b]	Total annual costs[c]	Capital investment	Operation and maintenance costs	Capital costs[b]	Total annual costs[c]
Air pollution										
Public	0.1	0.1	0.2	0.6	0.2	0.8	1.9	4.2	1.8	6.0
Private										
Mobile	3.4	1.5	4.9	1.3	4.4	5.7	29.0	20.9	31.7	52.6
Industrial	2.1	2.4	4.5	4.9	5.6	10.5	24.0	34.8	39.5	74.3
Utilities	1.0	1.0	2.0	3.5	3.5	7.0	19.0	21.0	21.0	42.0
Subtotal	6.6	5.0	11.6	10.3	13.7	24.0	74.0	80.9	94.0	174.9
Water pollution										
Public										
Federal	0.	<0.05	0.2	0.2	0.1	0.3	2.0	2.1	0.4	2.5
State and local	2.3	7.6	9.9	6.2	15.1	21.3	56.9	41.7	113.7	155.4
Private										
Industrial	1.9	1.7	3.6	7.4	5.7	13.1	41.4	40.6	33.0	73.6
Utilities	0.6	0.4	1.0	1.4	0.8	2.2	3.8	10.5	6.3	16.8
Subtotal	5.0	9.7	14.7	15.2	21.7	36.9	104.1	94.9	153.4	248.3
Radiation										
Nuclear power plants	<0.05	<0.05	<0.05	<0.05	<0.05	<0.05	0.2	0.1	<0.05	0.2
Solid waste										
Public	1.5	0.3	1.8	2.1	0.6	2.7	4.6	18.2	4.2	22.4
Private	2.9	0.7	3.6	3.9	1.1	5.0	5.6	28.4	8.8	37.2
Subtotal	4.4	1.0	5.4	6.0	1.7	7.7	10.2	46.6	13.0	59.6
Land reclamation										
Surface mining[d]	NA	NA	NA	0.3	0.3	0.6	1.7	1.8	1.6	3.4
Total	16.0	15.7	31.7	31.8	37.4	69.2	190.0	224.2	262.0	486.0

[a]Incremental costs are expenditures made pursuant to federal environmental legislation beyond those that would have been made in the absence of this legislation.
[b]Interest and depreciation.
[c]Operation and maintenance plus capital costs.
[d]Not included in this year's estimate.
NA = not available.

With respect to the goal of price stability, for example, the CEQ estimates[18] that the Consumer Price Index is presently 1.6 percent higher than it would be if no measures of environmental protection were implemented. This figure is expected to rise to 4 percent by 1983. On the average this means a loss in price stability of .3–.4 percent per year. We have already discussed the drain of resources from production for environmental policy measures, thereby causing a loss of 2.2 percent in the GNP. It is especially difficult to estimate the effect of environmental policy measures on the goal of full employment. On the one hand, such measures cause cutbacks in environmentally degrading sectors and mirgation of factors of production from these sectors. On the other hand, environmental policy measures also create new areas of demand. The CEQ is of the opinion that the overall effect of environmental policy on employment is positive when unemployment is at a relatively high level; when unemployment is reduced, the positive impact of environmental policies also shrinks. In discussing the opportunity costs of environmental policy with respect to major societal objectives, it must be remembered that certain sectors are affected more severely than others. Those in basic industries which are involved in processing raw materials and are energy-intensive, such as pulp and paper, metal finishing, iron and steel, electric utilities, chemicals, mineral oil extraction and refinement, are usually hit harder than other industries.

Table 6 provides an overview of the investments undertaken for pollution abatement in the United States[19] in 1975, differentiated according to air and water pollution, solid wastes, and economic sectors. Figures for pollution control as a percentage of total investments are also given. The table indicates that certain areas, such as primary metals (17.2), stone, clay and glass (14.3), paper (16.8), chemicals (10.9) and petroleum (11.8) achieve a high percentage of investments for pollution abatement.[20]

ENVIRONMENTAL QUALITY AS A TARGET OF ECONOMIC POLICY

Diagram 11 illustrates how the problem of environmental disruption can be integrated into economic policy. For the sake of simplification, the environmental system and the economic system are combined as the techological-economic system. This system determines the quality of the environment through the amount of emissions. The gross pollution de-

Table 6.[19] Investments for Pollution Abatement by U.S. Industries in 1975 (in millions of dollars)

Industry	Reported expenditures: 1975				Pollution control as a percentage of total plant and equipment investment		
	Total	Air	Water	Solid waste	Actual 1974	Actual 1975	Planned 1976
All industries	6,549	3,790	2,362	396	5.0	5.8	6.1
Manufacturing	4,475	2,494	1,736	245	8.0	9.3	8.9
Durable goods	1,775	1,161	529	85	7.3	8.1	7.9
Primary metals	1,012	750	221	41	16.6	17.2	17.3
Blast furnace steel works	396	261	135	1	12.1	13.5	18.8
Nonferrous metals	546	425	82	39	21.8	24.1	18.8
Electrical machinery	136	34	93	9	6.8	5.8	6.6
Machinery, except electrical	83	40	37	6	1.8	1.8	2.2
Transportation equipment	116	51	50	15	3.7	3.4	4.1
Motor vehicles	86	35	38	13	4.1	3.9	4.8
Aircraft	26	14	11	1	2.9	2.8	2.8
Stone, clay, and glass	198	164	31	3	12.9	14.3	11.5
Other durables	229	122	97	10	4.5	5.3	4.3
Nondurable goods	2,700	1,333	1,208	160	8.7	10.3	9.6
Food including beverage	175	71	92	12	4.7	5.2	5.4
Textiles	31	15	15	1	3.3	4.6	5.8
Paper	489	273	139	27	19.3	16.8	15.0
Chemicals	684	250	394	40	0.3	10.9	11.5
Petroleum	1,239	684	483	72	10.1	11.8	9.8
Rubber	41	25	14	2	3.2	4.0	4.6
Other nondurables	41	14	22	6	1.8	2.8	2.6
Nonmanufacturing	2,074	1,296	626	152	3.0	3.2	4.1
Mining	73	32	31	10	1.8	1.9	2.7
Railroad	35	11	21	3	2.2	1.4	1.5
Air transportation	11	6	4	1	0.7	0.6	1.1
Other transportation	41	12	19	10	2.3	1.4	2.1
Public utilities	1,700	1,138	466	96	7.9	8.4	10.1
Electric	1,650	1,123	438	89	8.9	9.7	11.9
Gas and other	50	16	28	6	1.5	1.5	1.1
Communication, commercial, and other	214	97	84	33	0.6	0.6	0.6

pends on production and consumption technology, abatement technology, sectoral economic structure, and the level of economic activity (GNP). The amount of pollution eliminated is determined by abatement technology and the adaptive capacity of the economic system to environmental policy measures. Gross pollution and emissions abated or avoided define the net pollution which, in conjunction with complex ecological interdependencies, determines the quality of the environment.

The political system compares the environmental quality which results from the technological-economic system with the desired level of environmental quality. The determination of the global variable "environmental quality" must take into account the demand for this public good (e.g., through referendums and elections) and the costs of its realization. The costs are an input into the political system while the environmental policy measures are an input into the technological-economic system. The costs of environmental protection include the costs of forfeited production and the opportunity costs with respect to other economic goals (full employment, price stability, and balance of payments).

STANDARDS AND PRICES APPROACH

In many cases it is impossible to assess environmental quality and determine the benefits. The cost-benefit analysis can therefore not be used. In this situation the level of environmental quality is set as a fixed target by a political decision-making process whereby the standard is combined with a price for emissions. This method is called the standards and prices

Diagram 11. Schematic Representation of the Standards and Prices Approach

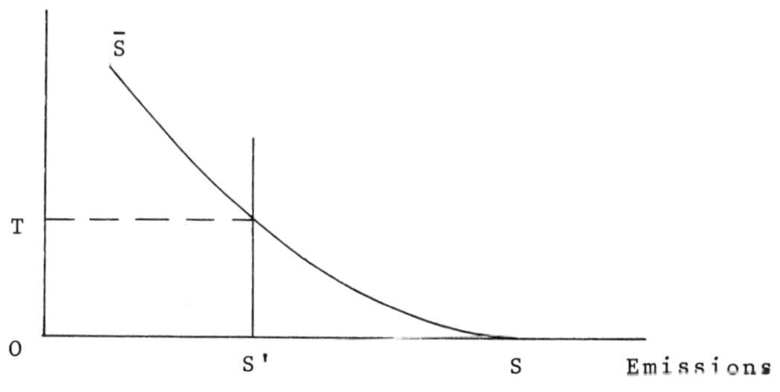

Diagram 11 Integration of Environmental Quality Considerations into Economic Policy

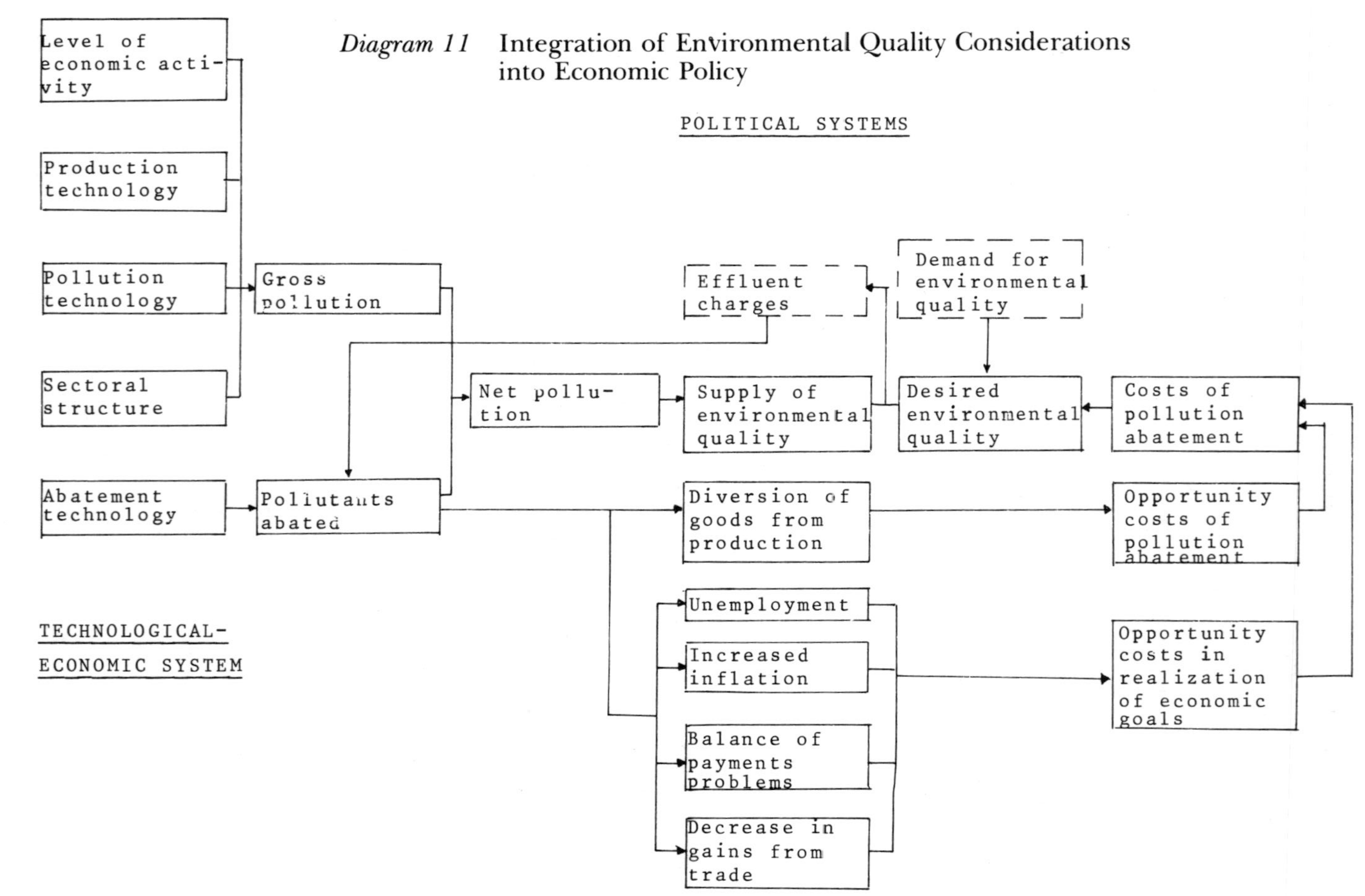

approach.[21] The economic dimension of the allocative problem is reduced to the question of achieving the desired level of environmental quality with a minimum of abatement costs. The situation is illustrated in Diagram 12. OS represents the point of departure of a given level of emissions. OS′ indicates the environmental standard or the tolerable level of emissions. SS′ denotes the quantity of emissions to be reduced. The curve $S\bar{S}$ shows the marginal costs of abatement. In order to achieve OS′, and effluent charge must be set at OT.

The standards and prices approach does not attempt to base emission charges on a determination of the benefits of environmental policy. It represents an attempt to achieve an environmental standard with minimal costs of abatement. Since the environmental standard is politically determined, the resulting level of environmental quality may be suboptimal. Only by chance does the standard set through political decision making coincide with the optimal level of environmental quality.

THE TIME HORIZON

The costs and benefits of environmental protection measures become apparent in very different periods. A water purification plant, for example, involves investments in the first period which are followed by maintenance and repair costs, and these can vary from period to period. On the other hand, the benefits can also be distributed unevenly over time, depending on the damage function. If the damages are caused by the pollutants released in one period only, then a measure to decrease the pollution will have a swift effect. If the damages stem from accumulated pollutants, such as DDT, then measures to reduce the emission of the pollutants will take ten to twenty years to reduce the presence of these pollutants in the environment.

For which time horizon should we plan? For a time period of ten, twenty, a hundred, or, as in the case of the earth's oxygen supply or the poisoning of the atmosphere, 10,000 years? The choice of the time horizon is largely a technical matter for individual projects such as water purification plants. The life span of the project must be calculated and the expected costs and benefits must be compared. But the problem takes on different dimensions when the issue boils down to "Environmental protection—yes or no?" Like damage assessment, the choice of a time horizon is in the final analysis a matter of values.

In order to understand the question of time horizons (a point already considered with respect to the planning models discussed in Chapter II), we propose a slight detour. Let us assume that a given measure for environmental protection provides a benefit of $50,000 in each period. Are the $50,000 of the first period worth the same as those returning twenty years later? Disregarding for now the element of inflation, won't the $50,000 be more important to you today? If you were promised better air now or in twenty years' time, wouldn't you prefer to have it today?

Economists refer to this phenomenon as the law of time preference. This means that a benefit in the present will be preferred to an equal one at a later date. If this is true, and if the benefits vary in each period, then they can be compared only if the benefits in the future can be corrected, i.e., reduced in value as a function of the societal time preference. The benefit of an environmental protection measure producing results in twenty years is of less value today because of the time preference factor. In order to obtain the real value of this measure, the future benefit must be reduced in accordance with a given societal time preference rate. Just as the sum of money which is lent today is less than the sum to be paid back in ten years, the accounts of future benefits must also be adjusted to changing values.

The choice of time preference rates is closely related to the problem of the time horizon. Table 7 presents the real value of a benefit from $100 in each period as calculated at interest rates of 4 percent and 8 percent. It indicates that a high rate of interest reduces the present value of fu-

Table 7. Discount Rate and Present Value of Benefit

Year	Benefit in each year	Present value of benefit at discount rate of 8%	4%
5	$100	$68.06	$82.19
10	100	46.32	67.55
15	100	31.53	55.53
20	100	22.47	45.64
25	100	14.60	37.52
30	100	9.94	30.83
40	100	4.60	20.83
50	100	2.13	14.07
60	100	0.99	9.51
70	100	0.46	6.42
80	100	0.21	4.34
100	100	0.05	1.98

ture periods greatly. In other words, high time preferences cause the benefits occurring in future periods to be increasingly neglected. High interest rates, then, imply a limited time horizon, and vice versa.

If one has a societal time preference rate of 8 percent corresponding to the capital market interest rate in a society accustomed to inflation, then the real value of a benefit flow of $100 which will occur in five years will be only $68, and that occurring in twenty years will be worth only $22. The benefit flow expected in the 60th period is so greatly devalued that it hardly has any effect on the present value of the project.

If, on the other hand, one chooses the lower societal time preference rate of 4 percent, then the future utility flows will be significantly less discounted and they will therefore lead to higher present values. While the utility flow in the 30th period has a present value of only $9.94 at a discount rate of 8 percent, at an interest rate of 4 percent the real value is $30.83.

When the real value is very small, the planning horizon has been reached, and the benefit flows accruing after this point no longer matter, if, for example, it is accepted that the planning horizon is reached when the real value of a utility flow of $100 is only $1, then the horizon will be attained in 50 years at a discount rate of 8 percent and in 100 years at a societal discount rate of 4 percent, again illustrating that a high interest rate means a shorter planning period.

The question as to whether a society should use a long-term or a relatively short-term strategy in its planning for environmental measures is, therefore, identical to the question as to whether a high or a low time preference rate is to be used. If we choose a high interest rate we give more weight to the present costs and benefits, i.e., to the present generation. A low interest rate gives greater weight to future generations. The choice of the discount rate consequently determines intergeneration equity.

Pigou believed that, in contrast to an individual whose life span is limited and thus tends to a high time preference rate, society should assume lower preference rates. On the other hand, it could be argued that as a result of technical progress, future generations will dispose of far greater wealth, which will facilitate investments not undertaken today. This thesis would lead to the choice of a higher interest rate.

There can be no doubt, however, that health damages appearing only in a distant generation, gradual environmental deterioration and long-term ecological changes all demand very long-term perspectives.

Further, benefits resulting from the avoidance of these damages should be handled with a very long time horizon, too. In this matter we fully agree with Joehr:

> In environmental economic terms we must behave in such a way as to permit future generations to live a life without external need. We may not deprive these generations of the possibilities of development which we have today. . . . This means an increase in the awareness of our responsibilities to future humanity similar to that experienced by people in the 19th century toward the underprivileged in their own country, and that felt by industrialized countries since World War Two towards the populations of developing countries.[22]

In addition to choosing the "right" time horizon, and deciding on very long-term perspectives for many environmental damages, we must take into consideration that values can change significantly over time. It is to be expected that the demand for many goods, such as drinking water, air, and leisure opportunities will rise steeply. A growing world population will have a major impact on the rise in demand. Another factor will be, as discussed above, that public goods will gain in value as wealth and private goods increase. The value accorded to environmental conservation will therefore automatically rise over time and the benefit flows of an improved environment will increase in the future.

Heavy emphasis on long-term perspectives in environmental policy making will increase the importance of asymmetry between the uses of the environment (Chapter IV), as some decisions are reversible and others are not.

Finally, distributional considerations are not usually included in cost-benefit analyses, since a given income distribution is assumed. This means that the impacts of environmental measures on the distribution of income are not taken into account in this assessment process. The next chapter will concentrate on this important point.

NOTES

1. B. S. Frey, *Umweltoekonomie,* Goettingen, 1972, p. 55.
2. See I. L. Knetsch and R. K. Davis, "Comparisons of Methods for Recreation Evaluation," in R. and N. Dorfman (eds.), *Economics of the Environment,* 1972.
3. P. Bohm, "An Approach to the Problem of Estimating the Demand for Public Goods," *Swedish Journal of Economics* 73:55–66 (1971).
4. W. C. Birdsall, "A Study in the Demand for Public Goods," in R. A. Musgrave (ed.), *Essay in Fiscal Federalism,* Washington, D.C., 1965.

5. R. G. Ridker, "Economic Costs of Air Pollution," *Studies in Measurement*, New York, 1967.

6. Council on Environmental Quality, 1971, p. 106.

7. T. A. Crocker, "Externalities, Property Rights and Transaction Costs: An Empirical Study," *Journal of Law and Economics* 14:451–464, (1971).

8. B. S. Frey, *Moderne politische Oekonomie*, Munich, 1977; A. Downs, *An Economic Theory of Democracy*, New York, 1975.

9. R. H. Coase, "The Problem of Social Cost," *Journal of Law and Economics*, 3:1–14 (1960).

10. J. H. Dales, *Pollution, Property, and Prices*, Toronto, 1968.

11. E. Furobotn and S. Pejovich, "Property Rights and Economic Theory, A Survey of Recent Literature," *Journal of Economic Literature* 10:1137–1162 (1972).

12. See T. N. Tideman and G. Tullock, "A New and Superior Process for Making Social Choices." *Journal of Political Economy* 84:1145 (1976).

13. As described by E. H. Clarke in *Multiparty Pricing of Public Goods, Public Choice*, Vol. II, 1971, pp. 17–33.

14. R. Pethig, *Umweltallokation mit Emissionssteuern*, Tuebingen (forthcoming).

15. Source: Council on Environmental Quality, 1971, p. 119.

16. Source: Council on Environmental Quality, 1976, p. 167.

17. Ibid., p. 150.

18. Ibid., p. 147.

19. Ibid., p. 166, from the U.S. Department of Commerce, Bureau of Economic Analysis.

20. For an international comparison of pollution control costs, see Table 8 in Chapter X.

21. W. J. Baumol and W. E. Oates, "The Use of Standards and Prices for the Protection of the Environment," *The Swedish Journal of Economics* :42–54 (1971).

22. W. A. Joehr, "Bedrohte Welt, Die Nationaloekonomie vor neuen Aufgaben," in M. P. van Walterskirchen (ed.), *Umweltschutz und Wirtschaftswachstum*, Munich, 1972, p. 85.

Chapter VIII

The Recipients of Public Bads

Of two evils, no one will choose the greater if he can choose the lesser.

Plato

Environmental quality is a public good. In contrast to private goods, it is characterized by the following traits:

1. According to typical definitions of a public good, many people can enjoy it simultaneously and it can be used to the same extent by all. Everyone can enjoy the beauty of a redwood forest, and nobody can be restricted from hiking in the Alps or swimming in the Atlantic.

Although the term "nonexclusive" consumption does eliminate some goods such as food, clothing, and housing, it is not a sufficient criterion for defining a public good. Collective consumption includes such activities as a boxing match, when many spectators can enjoy the activity at the same time and even add to the fun by clapping and shouting. But a boxing match is a far cry from a public good, so another term must be added to the definition.

2. No one can or should be excluded from using a public good, while the utilization of a private good can be refused to an individual unwilling to pay the price. The technical impossibility of excluding someone, illustrated by the example of the lighthouse, is rarely so clear-cut. Fences and gates can be constructed at public events, such as sports, to exclude those who do not want to pay. Cable television is another example of technical exclusion, as are toll roads, electric meters, special receiving equipment (i.e., radar), water meters, bouncers, and legal restrictions on use. Similar measures also permit the technical exclusion of individuals from con-

suming environmental services. Therefore, the deciding factor can no longer be the possibility of *technical* exclusion, but rather the social undesirability of exclusion. The importance placed by society on the right of all individuals to walk in the woods is the decisive element in the existence of this public good.

THE PUBLIC BAD

A climber on Mt. Whitney or a swimmer in Lake Erie are clearly consuming a public *good*. But what about the polluted air of Los Angeles? Obviously, the air is breathed by many individuals simultaneously; it is inhaled by millions. It is a basic consumer good and no one can be excluded from using it. But what is so "good" about this public good, in which carbon monoxide and oxides of nitrogen, sulfur dioxide and hydrocarbons are floating around along with many other corrosive or toxic substances which enter the lungs with each breath? What is so "good" about the public good "a swim in the Rhine," which is so polluted that although no one would be prevented from doing so, no one would even think of swimming in it today? Are these examples of a public good? Of something desirable and worth aspiring to? It would be more to the point to speak of public *bads* in such cases. This negative input into our consumption must be taken into consideration. It appears to be the other side of the coin of the public good. How can it be defined?

1. The public bad is usually borne by many individuals simultaneously. The collective use of a public good thus corresponds to the collective endurance of a given negative environmental quality. Many people must suffer the consequences of a weather inversion in Pasadena. It might be a comfort to know that Mr. Smith is suffering from the same carbon monoxide, but that does not reduce the amount we have to inhale.

2. No one can avoid a public bad. Just as no one can be excluded from using a public good, no one can escape the pollution of our drinking water, the deteriorating quality of our air, and the consequences of the rising carbon dioxide content of the atmosphere.

Definitions such as these are classifications of ideal types which do not always correspond to reality, and which do not satisfactorily account for actual phenomena, There are always border cases in which one or another of the criteria are not fully met. Can the environment as a consumer good really by used by all in the same way, or do some people benefit from better opportunities? Is the good really so *public* everywhere? And what about public bads? Is everyone exposed to the same amount, or can some people avoid polluted air more easily than others? Are exceptions so frequent as to challenge the validity of our definitions of "nonexclusion" for a public good and "unavoidability" for a public bad?

DISTRIBUTIONAL EFFECTS OF PUBLIC BADS

This is precisely our thesis: that a public good, contrary to its definition, cannot be used by everybody. There are many exclusion factors which permit differentiation in consumption. And similarly, a public bad must not be endured by everyone to the same extent because there are ways of avoiding the effects of public bads. The same determinants which define the distribution of the use of a public good also explain the differences between the endurance of a public bad. What are these factors?[1]

1. The definition of a public good is based on the assumption that this good can be used by many people and that the consumption by individual A is not affected by the use of individual B. This supposes that the public good in question has no capacity limits. However, our discussion of the competitive uses of the environment (Chapter IV) has revealed the fallacy of this assumption. If many people compete for the use of a public good, then the character of the good is lost. Certainly, people can continue to use it, but its quality decreases. When too many people swim in a lake (the single-purpose illustration of competitive uses) the benefit derived by each person from the consumption of the good is reduced. A public good is degraded through excessive use. It is therefore no longer possible to accept the definition of a public good as one which everyone can use without affecting the use of the good by other people.

The multipurpose variant, whereby one kind of activity affects another, also detracts from the quality of the public good. If a river is used to dispose of waste products, then the quality of the drinking water will be reduced. If pollutants are emitted into the air without any restrictions, the air we breathe will become polluted and we will no longer be able to have the fresh air we used to enjoy. The definition of the environment as a public good therefore presupposes that there are no competing uses, be it among many individuals, among alternative types of uses, among different production activities, among conflicting consumption processes, or between consumption and production. But this is an unrealistic premise, and the public good clause, "to be used by everyone without affecting others," has long since ceased to be applicable to environmental services.

If this thesis about the competing uses is correct, then it must also be true that public bads are endured in very different ways. Take for example the case of the competitive uses between the people having to endure polluted drinking water because they live downstream from a paper mill and those living on a lake which they themselves are polluting. In a case of "self-pollution" the criterion of passive endurance of the public bad is certainly not fulfilled in the same way as when people have no way of affecting the amount of pollutants emitted into their environment.

2. Theoretically, if the conflict between competitive uses of the environment has not been resolved, it is possible for people to use the environment any way they please. In fact, however, certain activities have such a powerful effect on the environment that they can dominate or even partly exclude other competing uses. The frequency and intensity of use are also important factors which influence a differentiation of the demand for the environment.

An analogous situation exists in the case of public bads. There are groups who suffer more than others from air pollution: heart patients, people with circulatory problems, and those with respiratory ailments. It appears that, as a consequence of varying intensities of demand, the public good is enjoyed differently and the public bad is endured differently.

3. The spatial delimitation of the public good and the public bad can also play a role. Many positive and negative environmental qualities are spatially limited, such as the beauty of a landscape, the ugliness of an industrial center, the pollutant content of the air, or the danger of toxicity of drinking water. The use of such public goods can be avoided temporarily, during vacations, or permanently, by changing jobs or place of residence. Public bads can likewise be avoided. The opportunities for use or avoidance are very different for different individuals.

This differentiation factor would cease to play a role if the public good or bad were spatially unlimited, such as a general lack of oxygen throughout the world, or an evenly distributed increase of carbon dioxide in the earth's atmosphere.

4. Spatial proximity is yet another determining factor in the enjoyment of environmental goods or the endurance of bad environmental

qualities. A person having the good fortune to live in a beautiful area with little pollution can make greater use of this public good than the inhabitant of a big industrial city.

Certain costs must be paid in order to be able to enjoy a lovely landscape or to avoid a negative environmental condition. The decision to move involves sacrificing a present job and home and searching for new work and housing, together with all the attendant expenses. There are also the costs of transportation to enjoy environmental services for a short period of time, as well as the transportation costs to avoid a negative environment, such as when individuals choose a residence far from the polluted air of the city in which they work. All costs of this nature limit access to positive environmental quality.

5. Differences in the use of a public good and the endurance of a public bad depend on the level of information available to the individual. People who do not know the pollutant content of the air at their place of residence are in no position to alter their behavior accordingly. In certain U.S. cities pollution levels are reported on television and in the press along with the weather. However, this information is rarely provided in other countries.

6. Private goods are another factor in the use of certain environmental qualities. Traffic jams and other transportation problems limit access to the wilderness on the weekends. People who have a swimming pool do not have to rely on a public pool. And individuals who have air conditioners with filters in their homes can control the air they breathe to a certain extent. The acquisition of land and the ability to exclude others also have some importance for the possibility of enjoying the amenities of the environment. Large sections of the population can be excluded from using a lake shore or a beach because they cannot afford the high cost of land, housing, or rent in the area.

7. Population density also is a factor which differentiates the use of the public good. If the demand for the environment is too great in a heavily populated area, a given service can become scarce even if it is being used only for a single purpose. Through excessive demand the capacity limit of the public good is reached and its quality changes. Individuals living in less heavily populated areas can make more use of similar goods. Since population and industrial density create more pollution, the inhabitants of heavily agglomerated areas have to endure more public bads.

8. Finally, a differentiation of use can occur if exclusion mechanisms

are introduced when capacity limits are reached, ranging from first-come-first-served measures to allocation according to societal priorities.

In summary, many factors lead to an unequal distribution of the benefits of enviromental services and the liabilities of the public bads. Public goods and public bads are not as "public" as theory seems to indicate.

These theoretical suggestions have been substantiated by empirical studies which show that the consumption of environmental services varies positively with income and wealth. For example, a systematic inverse relationship between income and environmental quality was observed by a study conducted in Kansas City, St. Louis, and Washington, D.C.[2] Although there are differences between the cities investigated (e.g., the lowest income group in Washington, D.C. suffers on the average from a lower level of air pollution than the highest income group in St. Louis); the distribution of damages within each city is weighted toward the lower-income groups.

The difference between damages suffered by whites and blacks was particularly stark. In each of the cities the black families were exposed to higher concentrations of the pollutants tested (particulates and sulfur trioxide) than the average family with an income lower than $3,000 per year. In a survey[3] of New York the air quality was tested at 38 control stations. Curves for similar air quality were constructed to produce a picture of the spatial distribution of air pollution as measured by four indicators of air quality. The income distribution within the city was also ascertained, and expressed in percentages of low-income groups in each area. Correlation analyses and a comparison of the spatial distribution of air quality and the population indicated that lower income groups are exposed to relatively higher levels of air pollution.

THE NEW DISTRIBUTION PROBLEM

If public goods and public bads are taken into account, then income distribution can no longer be defined simply as the distribution of the income earned by individuals in one period. Income also consists of the utility flow derived by an individual from the consumption of various goods. These include not only private products, but also public goods, such as the environmental services, and public bads.

This fact places the distribution problem in a very different light. The question is no longer how the national income will be distributed among individuals or factors of production within one economic period, but rather how the goods available for consumption in a given period are distributed among the population. This very broad interpretation of the distribution problem will, when it has been fully elaborated, lead to a revision of the principles of distributional theory and require new approaches in distributional policy. The central variable of distribution theory is significantly expanded when considerations of utility differentiations of public goods and public bads are included, and experts will have to revise objectives and policy measures in the light of this.

The recognition of the distributional aspect of public goods and public bads has three major consequences for our discussion of environmental problems.

1. Neither planning models (Chapter VI) nor cost-benefit analyses (Chapter VII) take into consideration the distributional aspect as it is discussed here. The distribution of the benefits and costs of environmental protection measures as presented in cost-benefit analyses are usually forgotten. At best they are accounted for in an ancillary calculation. But these distributional effects have an impact on social well-being and are, therefore, an important aspect of cost-benefit analyses. Such analyses are indeed intended to provide information about changes in societal well-being, about the social groups which benefit from environmental measures, and about those who must bear the costs.

2. Damage assessment is gauged by the individual's willingness to pay for reducing environmental pollution. Since the amount an individual is prepared to pay is a factor of the person's income, income distribution becomes important in the assessment of damages. The damages are overestimated if a high-income group is surveyed, since it can be assumed that individuals having a higher income set a higher priority on a clean environment. The identification of utilitiy for given or proxy market prices also depends on certain demand relationships, which vary with income distribution. The situation is analogous with respect to the distribution of wealth.

Since individuals with high incomes accord a relatively high value to environmental quality, it is often argued that this group in particular benefits from environmental measures. However, this view contradicts the assertion that environmental bads are especially damaging to lower-income groups. The same argument can surely not be used to support two opposing hypotheses. If the lower-income groups suffer the most from air pollution, then they must also benefit the most from pollution abatement.

This contradiction resolves itself when it becomes clear that the benefits are measured by different variables in the two cases. The fact that lower-income groups are exposed to more air pollution refers to the amount of pollutants in the air. The benefits of environmental protection measures to these income groups consists in the reduction of pollutants. The benefit of environmental improvement to the higher income groups, however, is measured by the value placed by these groups on environmental quality, not according to the amounts of pollutants. It is obvious that the individual from a higher income group will place more value on the reduction of pollutants.

The competition for program funding is another area revealing the relatively disadvantaged position of low-income groups. Minority organizations in the United States emphasize that expensive environmental programs drain resources from important social programs like urban renewal, which benefit lower-income groups.

Finally, one more distributional aspect of environmental policy must be considered, namely, the distributional effects of different approaches to financing environmental policy measures. For example, certain measures, such as charges, function regressively; that is, low-income groups bear a relatively greater portion of the burden. Subsidies, on the other hand, may be less regressive. It must be observed, however, that environmental protection is not primarily a distributional but rather an allocative problem.

3. The question of distribution also has an international aspect. The developing countries argue that environmental policies are a luxury of the wealthy, developed nations. In view of the poverty of the Third World the growth of the GNP must clearly be accorded priority over environmental measures. These countries fear that the funds now used for environmental purposes in developed countries will siphon off money from development assistance funds. They suspect that the allocation of investment credits is tied to environmental expenditures, which will have significant effects on economic growth in developing countries. They also worry that production standards will create new trade barriers for their exports, the revenue from which they need desperately, and that the implementation of such production standards will involve heavy costs, thereby eroding their competitiveness and delaying their development. They fear, lastly, that all measures for recycling material in the developed countries will lead to sparing use of resources and thereby reduce the demand for their export articles.[4] (See Chapter IX.)

According to the developing countries, then, the wealthy nations which are generously endowed with private goods would profit from stringent environmental policies because they can afford to place a high value on environmental quality. These environmental regulations, however, would limit the economic growth and foreign trade of developing countries.

The developing countries provide not only raw materials and similar inputs to production and consumption processes in the developed countries, but also many public goods. Returning to our discussion about spatially limited goods,[5] such as landscapes, it is clear that therein lies an opportunity for the developing countries to benefit from the disadvantages of industrialization in the developed nations. The developing countries can offer such goods to developed nations in the form of tourist attractions, exercising the exclusion principle characteristic of such good. The more industrialized the developed nations become and the worse the degradation of the environment, the higher the demand will become for the environmental services which the developing countries have at their disposal. This distribution of regional goods can become a significant source of income for the Third World.

The problem of global public goods is more complex. The country of origin cannot exercise the exclusion principle because the good is not spatially defined. For example, Brazil's virgin forests produce oxygen for the world, but the country cannot demand payment from another nation which is unwilling to pay. The United States, on the other hand, uses more oxygen than it produces. The poor country produces a good which the rich nation can use free of charge. Only if property laws could be internationally agreed upon and fees established for the consumers of the goods could the international distribution problem of global public goods be resolved.

EQUITY VERSUS EFFICIENCY

In this chapter the distributional aspects of environmental policy have been emphasized. It is clear, however, that economic problems and environmental policy have other aspects, the primary one being allocative. Environmental problems appear because private and societal costs diverge. This allocation aspect must not be forgotten in attempts to resolve the problem of distribution. As much as the goal of equal distribution of income might correspond to our sense of fairness, it cannot be denied that too great an emphasis on this point can actually reduce well-being. Policy measures to equalize the distribution of income unfortunately sacrifice efficiency, and thereby the goods available for consumption. If the pie is divided equally among all members of society, the pie will shrink because distribution influences incentives for supplying goods and factors of production. For example, the emphasis placed in Great Britain on equitable distribution in the 1960s and 1970s had a negative impact on the GNP. In other words, there is no point in guaranteeing everybody the same size piece of pie if this measure makes each piece smaller.

If too much weight is accorded to equitable distribution, there will be great resistance to possible measures for the improvement of environmental quality, and the decisive allocative aspect of the problem will be submerged. Social groups must be prepared to calculate into their real income the value of an improved environment, and must be willing to sacrifice income increases in the traditional sense of the term in favor of improvements in environmental quality.

NOTES

1. For example, see H. Siebert, "Zur Frage der Distributionswirkungen oeffentlicher Infrastrukturinvestitionen," in R. Jochimsen and U. E. Simonis (eds.), *Theorie und Praxis der Infrastrukturpolitik,* Berlin, 1970, pp. 33–71.
2. A. M. Freeman III, "The Distribution of Environmental Quality," in A. V. Kneese and B. T. Bower (eds.), *Environmental Quality Analysis: Method and Theory in the Social Sciences,* Baltimore, 1972, pp.262 et seq.
3. J. M. Zupan, *The Distribution of Air Quality in the New York Region,* Baltimore, 1973.
4. H. Siebert, "Trade and Environment," in H. Giersch (ed.), *The International Division of Labor, Problems and Perspectives,* Tuebingen, 1974, pp. 108–121.
5. B. Frey, *Umweltoekonomie,* Goettingen, 1973, p. 83.

Chapter IX

Subsidies, Taxes, Standards: No Charge for Environmental Pollution?

The art of taxation consists in so plucking the goose as to obtain the largest amount of feathers with the least possible amount of hissing.

Ascribed to Jean Babtiste Colbert (1619–1683)

Diagnosis should be followed by therapy. After the causes of environmental degradation have been analyzed, and after their effects have been examined, economic policy measures for correcting the problem must be chosen. If the environment can be used for waste disposal at zero cost then the market cannot provide an adequate supply of environmental quality, and decision makers will have to intervene in the market process of the invisible hand. Crucial decisions are involved here. How should the institutional setting of the market system be changed? What measures are available? What solutions to environmental problems can the economists propose?

POLITICAL SWEET TALKING

Whenever the private sector does not behave according to the objectives of societal policy, the politician can try to persuade the business involved to conform. This "moral suasion" consists of appealing to businesses to accord more weight to environmental costs in decisions relating to production and investment. It also includes beseeching, imploring, and stimulating consumers to behave in a more environmentally sound manner, with the corollary warning that if the desired behavior is not forthcoming, more severe measures will be implemented.

The success of moral suasion in this as in other areas of economic policy (income policy guidelines, monetary and fiscal policy, energy conservation) depends on many conditions, including the politician's charisma and power of persuasion and the loyalty of the constituency. In societies which place heavy emphasis on individual freedom of choice, the politician's word is not a binding directive. Hence, there result severe limits to the power of persuasion.

This should not imply that political sweet-talking is totally useless. A major element of moral suasion is the provision of objective information to the public about the consequences of prevailing economic behavior. An optimist might believe that the awareness and discussion of environmental damages, the interest in the living space of humans, the recognition of the interdependence between the economy and the ecology, the discovery of pollution's destructive effects on health would change the personal value structures and thereby influence behavior patterns.

This is, however, an area of speculation, and realists will point to the fact that the best information about the dangers of smoking, the prohibition of cigarette advertisements on television in the United States, and the printing of health hazard warnings on cigarette packages have not resulted in a long-term drop in smoking. How much more difficult will it then be to use moral suasion for environmental policies when Mr. Smith pumps his pollution into other people's water and air rather than his own? Where the account is not drawn between the consumption and the damages of one party, but rather between the production or consumption of one party and the damages suffered by another? And where the damages caused by Mr. Smith distribute themselves among so many people that they are almost impossible to identify for one individual? When the environment is treated as a free good which can be used without restrictions and for the disposal of any pollutant? Moral suasion can certainly supplement other measures of environmental policy by explaining more severe measures. But, the "for-heaven's-sake-appeal" is a very blunt sword and one cannot seriously recommend reliance on this weapon to wage the war against pollution.

WHAT ABOUT SUBSIDIES?

Subsidies represent a second method. They are a favorite tool in government ecoomic policy. Nobody is visibly disadvantaged by them, and the party subsidized is full of praise. In nearly all countries of the Western world subsidies in explicit or hidden form play an important role. The agricultural sector is subsidized in all Western countries in one way or another. For instance, the European Common Market has developed a delicate system of political price-fixing for agricultural products. Quite a few sectors are supported by government subsidies, such as coal mining, ship building and airplane construction industries. Exports receive government aid and the sector of import substitutes, like shoe and textile industries in industrial countries, which are not quite competitive internationally are protected by import restrictions, tariffs, and administrative trade barriers.

If subsidies are so popular, then why can't they be applied for environmental protection? Isn't this a good cause? Won't industry be happy to accept this approach? Unfortunately, there are a number of reasons why subsidies are the worst possible approach to environmental protection. In order to understand the dangers of this method, we must first examine in detail the various forms that subsidies can take.

First, firms can be rewarded for installing waste-treatment equipment by receiving an advantageous tax credit, deferral or reduction from the Internal Revenue Service. The flaw inherent in this strategy is that it does not determine whether the investment was efficient in the economy as a whole, or whether the same amount might have been better invested in another firm. It is also not known to what extent environmental pollution is reduced by this measure; we only know that outlays for abatement have been made.

What if subsidies were related to the amount of pollutants abated? An answer to this question might be the anecdote about the Missouri farmer who asked his neighbor how much more he had not produced this year to increase his subsidy. This is the case which would be faced if subsidies were granted for the removal of pollutants of certain products. Just as the government pays for unproduced amounts of agricultural goods because of overproduction, it would reward firms for not producing pollution. To carry this example a little further, the government would then also have to pay a subsidy to a polluting firm which stopped business altogether so as not to pollute. A clever businessman would then devote himself to proving how much pollution he would have produced if his business were still operating, and how much production increase he sacrificed in order to reduce pollution.

Irony aside, subsidies for abated pollution raises serious problems. Such a measure would temporarily have a negative effect on the environment because firms would delay introducing environmental protection measures until the law came into effect so as to obtain the bonus. Secondly, the possibilities for abating pollution vary with different businesses. As a result, subsidies would provide a profit for firms in which the reduction of pollution costs less than the subsidy rate. And other firms would not introduce measures to reduce pollution if the costs were greater than the subsidy rate.

In addition to these problems, all forms of subsidies have the following important disadvantages:

—Subsidies do not guarantee that the quality of the environment will really improve. A subsidy may be awarded, for example, to a firm which introduces equipment to reduce the absolute amount of pollution created at a given level of production. But when the firm increases its production, the amounts of pollutants emitted will also rise. Since subsidies cannot be adjusted easily to handle this eventuality, pollution can increase in spite of them.

—Subsidies are always a means of retrospective correction. Such an approach is not satisfactory in a growing economy, where measures must be devised to impede environmental damages before they occur, rather than to try to correct them after considerable harm has been done.

—Subsidies must be financed through taxes which are a burden on the public. Should the taxpayer consider it strange to subsidize a business which damages the public, we are in complete sympathy.

—Subsidies promote the sale of environmentally damaging products. They cover a part of the social costs of the production of these goods. As a result, the business does not have to shift these costs to the prices, and environmentally damaging goods do not appear to be more expensive than their environmentally sound counterparts. The relative price structure does not change, so the price system as a allocative mechanism cannot function properly. Too many environmentally degrading products continue to be demanded and produced. The subsidy thereby achieves a devastating systematic distortion of the market economy and compounds the flaw in its construction discussed in Chapter IV.

The above observations illustrate the basic weaknesses of using subsidies to protect the environment.

The principle on which subsidies are based is "the victim pays." Yes, that's right: the victim pays. The victim pays the polluter to reduce wastes and pollutants. This type of arrangement can sometimes be worked out on a one-to-one basis to the satisfaction of both parties. If the incineration of autumn leaves by A is particularly bothersome to B, then B can propose to pay for the removal of the leaves to the local dump. The problem is not as easily solved in the case of subsidies from the government, when the taxpayer is not always identical to the victim. It somehow does not fit into our concept of fairness that C should pay A to eliminate the pollution that is bothering B.

In spite of all the objections raised here, subsidies are becoming increasingly popular in international relations. France expects to receive a subsidy from West Germany to clean up the pollution in the Rhine caused by the French potash plant in Alsace. The Netherlands want to dispose of untreated waste water in the Dollart River and would purify the water only if they receive a subsidy from West Germany. For its part, West Germany bestows pollution on the Netherlands by way of the Rhine, and might consider cleaning up the wastes if the Dutch pay.

THE VICTIM SPEAKS UP

Another approach to the problem entails establishing a right to a clean environment, whereby the victim could sue the polluter for infringing on this right. Such a principle could be applied both in or out of court. If, for example, the polluter can bribe the victim to accept the damages, then the polluter could continue degrading the environment and—in economic terms—neither party would suffer a loss of utility. Compensation payments would become a part of normal business expenses. Of course, paying compensations represents an incentive to reduce pollutant emissions.

As simple as it may seem, this approach to the allocation of environmental services has several disadvantages.[1] The theory assumes that there is only one polluter, and that both parties are equally strong. Usually, however, the parties are not equal in strength, since there is a difference in the amount of money they have available for the dispute. In addition, the victims are often in the majority, the total damage being measurable only as a sum of the many individual damages. This in effect reduces the interest of each individual in the dispute. If these victims face only one polluter, who is usually better organized and has more funds at his disposal for the case, then the bargaining positions are clearly unequal.

The theory also does not cover those cases in which there are many polluters. The smog of Los Angeles is produced by several million cars and by the energy consumption of eight million inhabitants as well as by industry. Under these circumstances it is extremely difficult to exact compensation from each polluter, particularly considering the varied diffusion processes in the environmental media.

In some cases it may appear desirable to deal in court with the conflict over the competing uses of the environment. If the victim cannot come to an agreement with the polluter in an informal bargaining process, he should have recourse to legal proceedings. Is this a realistic way of allocating environmental goods and services? Allocation mechanisms must a) function quickly, and b) be useful in decisions for future periods. The courts can fulfill neither of these two conditions. Nobody would consider allocating the factors of production in the economy through the courts. How many plaintiffs would starve by the time a decision was reached? Legal proceedings entail such high costs that they are at best a last resort.

The assumption that environmental conflicts can be boiled down to one polluter and one victim raises several legal problems. Who should bear the burden of proof? How can a polluter be sued for cumulative damages suffered by a large section of the population? This problem has been partially resolved in the United States with the establishment of "class action suits." With this approach it was possible for 2,000 prawn fishers to file a suit against Chevron Oil when an oil spill damaged the fishing industry in the Gulf of Mexico. However, this institution is not written into the legal statutes of all the states, and it is not implemented in all courts.[2] Nor has the option been introduced in other countries.

The right to a clean environment cannot be considered a primary measure in resolving the conflicts over the environment. At best it represents an emergency brake which citizens can pull when other allocative mechanisms have failed. This is not to repudiate efforts to firmly establish the legal basis of every citizen's right to a clean environment. Rather, it is meant to discount this legal tool as a key means for solving the complex problem of environmental pollution. Other measures and forms of organization must be found for this purpose.

THE BUREAUCRATIC APPROACH: EMISSION NORMS

Prescriptions and prohibitions are very appealing measures of environmental policy, particularly for ecologists. This is because these measures have the decisive advantage of guaranteeing a reduction of pollution—unless the polluters break the law. The common denominator of all prohibitions is that they contain direct stipulations about production and investment. Various types of regulations are presented below:

1. Emission norms usually establish in absolute terms the highest permissible amount of pollutants which a firm may discharge into the environment.

2. Obligations to reduce pollution, on the other hand, define a specific amount of pollution to be reduced, either in absolute terms or as a percentage of present emissions.

3. Technical specifications for production processes or equipment for pollution abatement and recycling force firms to achieve a particular level of technology. In some countries this instrument can be applied within the existing framework of investment controls in issuing permits for new production facilities. Regulations can also be applied to existing facilities.

4. The use of particularly polluting inputs should be regulated, whereby spatial and time differences must be considered.

5. A special type of emission norm is the specification of product norms. Here the quality of a product is defined by taking into account the pollution caused in the use of the good. Product norms can be implemented in both production and the consumption processes. Technical regulations on mass consumer goods such as cars, decibel levels for noise pollution, and the definition of polluting inputs such as sulfurous heating oil are examples of possible applications of this approach.

6. Limits can be set on the amounts of a good which may be produced. This tool can be used for specific periods of time and in certain areas to control products whose production process is particularly polluting in especially serious situations. A special form of this approach is total prohibition to produce a given product.

7. More stringent land use and construction planning, including more restrictive industrial siting regulations comprise yet another means available for environmental policy to control investment.

A key question in the implementation of environmental policy is how the desired standards of ambient environmental quality can be translated into emission norms for each individual firm. If a reduction in emissions is set in terms of percentage of present emissions for all firms, then those firms which have not yet implemented measures to reduce pollution have an advantage. This approach is arbitrary and inefficient.

Another possibility is to set emission norms according to the size of the firm, based on number of employees, capital, turnover, square footage, etc. However, this method prescribes a certain structure. It refers to a particular situation and therefore cannot satisfy the condition of dynamic allocation efficiency. Nor does it correspond to the concept of structural change which is desirable from a macroeconomic point of view. Specifically, this approach does not take into account the potential growth within a firm, thereby possibly placing small dynamic firms at a disadvantage. Emission norms based on fixed size of firms can cause distortions in the conditions of competition. They can restrict the entrance into the market of newcomers by prohibiting the establishment of new businesses when the assimilative capacity of the environment has been reached. It is then possible that a new firm which has up-to-date environmentally sound technology and causes less pollution per unit of value added than existing businesses is not permitted to locate in the area because of the fixed structure of location and infrastructure.

Air quality management in West Germany provides a good example.[3] The relevant law specifies the environmental quality for a region. If a new firm wants to locate in the region, it must prove that its technology corresponds to the "state of the art." The government agency responsible for issuing the location permit must prove that the firm does not meet the necessary criteria if it does not want to issue the permit. The system therefore does not contain any strong incentive to develop new technologies for pollution abatement. If the tolerable level of emissions has been exceeded in a region, a firm cannot locate there, even if its abatement equipment is highly efficient. The firms already established in the region, however, continue to emit pollutants. Such a solution of the environmental problem is conceived for a static economy and does not take into account the needs of a dynamic situation.

Diagram 13 illustrates why the reduction of emissions by X percent is not efficient.

The point OA indicates the emission of firm 1 and OA_2 those of firm 2 at the outset. Let A_1B_2 and A_2B_2 be the marginal cost curves for the abatement of the two firms. If the objective of the policy is to reduce the emissions of all firms by a third then firm 1 must abate A_1C_1 at cost $A_1C_1K_1$, and firm 2 must abate A_2C_2 at cost $A_2C_2K_2$. This is not efficient because the remaining pollution C_2C_1 can be abated at lower costs by setting an emission tax at OT. In this case, firm 1 abates the amount $A_1C'_1$, and firm 2 reduces the amount $A_2C'_2$. If the environmental policy maker wishes to achieve a certain level of environmental quality through emission norms at minimal costs, then a distinction must be drawn be-

Diagram 13. Application of Emission Standards

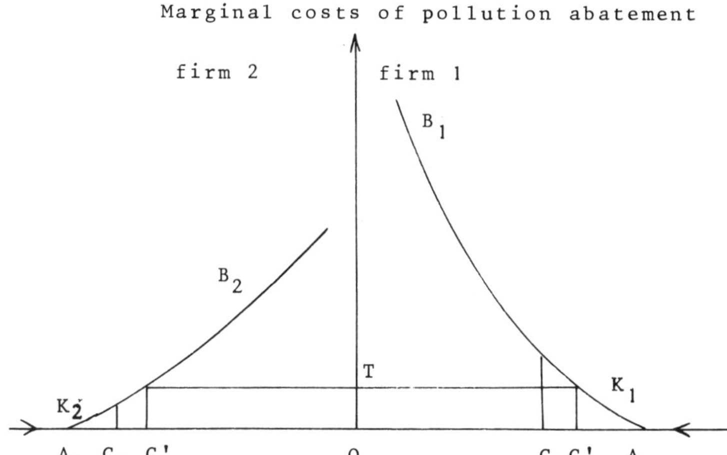

tween firms. Firms whose marginal costs of abatement are lower should be required to abate a greater percentage of pollutants.

Norms based on processes, inputs, or outputs are probably even more inefficient in economic terms than are general emission norms. While the latter leave the choice of inputs, outputs, and processes up to the individual firm, which usually tries to minimize the costs of production, the former approach restricts the latitude of the firm so greatly that, in view of the present low levels of information available to environmental agencies, economic inefficiency is certain to result. If it is avoided, then it is only by pure luck.

In addition to these problems, emission standards have several difficulties in common with other measures of environmental policy. Standards must be set for many pollutants and for various media. They also require knowledge about damage and diffusion functions. And since there are in fact a great many different damages, these too must be compared and related to a uniform measure. The equivalence of the different standards for the pollutants and the environmental media must also be monitored, to avoid the overexploitation of one if measures regulating others are stricter.

The major advantage of emission standards is that they do guarantee a response from industry. However, this certainty is obtained at the cost of heavy government intervention in private production and investment decisions.

TIME TO PRICE THE ENVIRONMENT

The use of the environment as if it were a free good is the decisive cause of its systematic misallocation in the market system. This flaw in the construction of the system favors the production of polluting products and places environmentally sound products at a relative disadvantage. As discussed in Chapter IV, it leads to the overuse of the environment and to a discrepancy between social and private costs. Obviously, the environment can no longer be used as a free good and must somehow be economically managed. An economic solution to the problem must therefore be based on a correction of the market system.

A possible way of regulating the use of the environment is to set a price on certain activities in the form of a tax or charge. For example, businesses or communities could be taxed for every ton of pollutants they discharge into the environment. This is called an emission tax. This would prevent everyone from using the environment for every imaginable purpose. A tax system is advantageous because:

—It guarantees that decentralized units discharging wastes into the environment will pay for this luxury and will therefore correct their economic behavior. Specifically, it introduces an incentive to reduce pollutants.

—If the tax corresponds to the social damages, then the discrepancy between social and private costs is eliminated.

—Products which degrade the environment will become more expensive, while environmentally sound goods will become relatively less expensive. The change in the price structure leads to a restructuring of consumer demand and thereby to a reallocation of the factors of production in favor of environmentally sound products.

—It encourages decentralized units to take advantage of improved, environmentally sound production inputs and processes made available by technical progress.

—It would provide an income which could be used for financing other environmental protection measures.

—It permits the maintenance of decentralized allocation and of the price mechanism. In comparison with other measures it can be implemented with relatively little government intervention, with the exception of setting the fees and controlling the pollutants discharged.

Clearly, such a tax cannot be a mere "token." Its impact must be felt by the polluter. It must be strong enough to actually change behavior by affecting production and investment decisions. One more point must also be clarified. A transition from a free environment to one that is economically managed through emission taxes involves a significant change in the data set of households and firms. If a price had been demanded in the past for the use of the environment, the tax would have increased gradually with industrialization, population growth and agglomeration. The change would not have appeared that drastic. But it is late now, and the longer we wait to correct the market system, the more severe the measures will have to be in the future.

Of course there are many arguments presented against a tax system for regulating environmental services. Only too often these criticisms are based on false preconceptions. Before jumping to conclusions, the following points should be considered:

—Charges and taxes do not, as many people seem to think, spell out the end of the market system. On the contrary, they introduce a desperately needed change and permit the maintenance of the allocative price mechanism.

—It is argued that products will become more expensive and that the consumer will have to bear the increased costs in the long run. Of course it is to be expected that environmentally degrading products will become more expensive for the consumer, but this is desirable and absolutely necessary. Such products must become perceptibly more expensive so that they can cause a change in the structure of consumption in favor of environmentally sound goods. If it turns out that businesses can mitigate this price effect by advertising environmentally damaging products more intensively, then policy makers must be prepared to severely restrict such advertising campaigns.

—The introduction of taxes and charges has been criticized because it will probably create friction in the labor market and require changes in the structure of employment. Obviously this will be the case. The development of environmentally damaging sectors will have to be curtailed and environmentally sound sectors will have to be promoted. A reallocation of the factors of production in favor of environmentally sound products will result.

—Finally, some countries fear that their products will no longer be competitive on the world market if environmental protection measures are implemented. Interestingly enough, both capitalist and socialist economists advance this thesis. For example, the East German economists Grundmann and Stabenow write:

> Any attempts by the GDR to resolve environmental problems are inexpedient because expenditures for this purpose are, in the first place, an economic burden which increases the costs of production and, if unilaterally introduced, decreases the competitiveness on the world market. In the long run this means a reduction in future capacity.[4]

Their argument is very questionable. First, some major countries have already introduced intensive environmental protection measures, so that the fear of one-sided increases in costs is unfounded. Secondly, should industries maintain competitiveness at the cost of a healthy environment for humans? What is the benefit of an additional export if its production degrades the living conditions in the exporting countries?[5]

HOW HIGH A TAX?

Clearly, the establishment of a price system for the environment raises many questions whose ramifications must be considered in order to provide an understanding of the proposal.[6]

1. Since the environment cannot be apportioned, an emission tax cannot be set per environmental unit used. It must therefore be expressed in terms of dollars per unit of pollution. An important prerequisite for establishing the rate is the exact calculation of emissions. If air pollution is to be reduced, for example, then there should not be a blanket tax on the input of coal, but on the amount of pollution resulting from the use of coal. If the tax is based on the amount of coal used, then the allocation of the factors of production will be affected even when coal is used in a nonpolluting process, such as in the chemicals industry. A tax based simply on the input of coal does provide an incentive to replace this energy source with others, but does not create the incentive to develop equipment that reduces the energy residuals.

Considering the lack of information, it might appear tempting to base taxes on the environmentally damaging product rather than on the amounts of pollutants. But this approach could cause problems. For instance, each time a product's pollution is reduced, the tax rate must be adjusted accordingly to reflect the new situation. But if political or administrative circumstances make the necessary rate changes unfeasible, then a direct relationship between the amount of pollution and the change no longer exists, and the incentive to reduce pollution disappears. At first glance, a tax on environmentally damaging inputs (sulfurous heating oil, for example) appears more promising. But the incentive function is distorted here, too, because this tax impels businesses to reduce the input rather than the pollution this element causes.

We therefore propose that charges be set on the actual emissions. In water management, steps in this direction have already been taken. The example of the water management cooperatives in the Ruhr area which have received world-wide attention[7] shows that pollutants can be measured for practical purposes. The costs of measuring air pollutants emitted by an industrial plant are not prohibitive for environmental policy purposes. For example, German studies indicate that CO, SO_X, HC, NO_X, and particulates can be measured with investment costs of approximately $100,000 per smokestack.[8]

2. Ideally, a tax set on emissions should reflect the damage caused by this amount of pollutant. Therefore, for this as for every environmental policy measure, information about the damage function is a vital starting point.

Diagram 14 shows how an emission tax is set. The curves indicate the necessary information: the marginal costs of disposal and the marginal

Diagram 14. Calculation of an Emission Tax

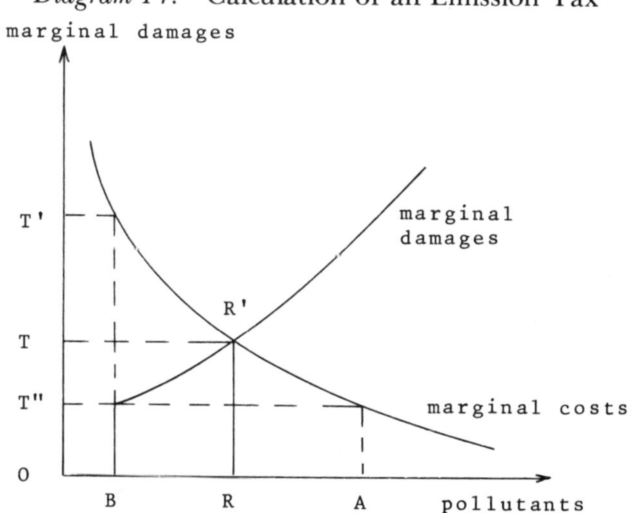

damages in relation to the amount of pollution. Marginal costs are the increase in the costs of disposal per additional unit of pollution, Marginal damages are the increase in overall damage which occurs with the addition of one unit of pollutant. The acceptable level of environmental pollution is defined by the point at which the overall societal damage is at a minimum, that is, when marginal costs equal marginal damages. This is indicated by point OR.

The rate of the tax must be designed in such a way as to achieve this optimal level. Environmental decision makers have three approaches at their disposal on which to base the rate of the tax.

Strategy I: Shock charge. The environmental agency orients itself to the damage function which indicates the societal benefit obtained from the removal of one unit of a pollutant. In this case the charge would be OT′. The benefit will be great for the removal of the first pollutants, as measured according to the marginal damage avoided. Thus, the tax will be very high. The economy will reduce a large quantity of pollutants with the amount OB remaining in the environment. Such a high tax, however, would cause the removal of more pollutants than is desirable. Consequently, the tax could be eliminated after a while. The disadvantage of this approach is that it sets in motion enormous adaptation processes such as changes in the structure of employment and in capital outlays for pollution abatement which also involve serious social costs. These changes must be corrected when the tax is reduced or eliminated.

Strategy II: The soft way. The environmental agency orients itself to the marginal costs of abatement for the polluter, rather than to the marginal benefits to consumers. A tax is set low at OT″ first, with a view to raising it in the future. Such a low tax would only induce minimal reductions in pollution, i.e., the quantity OA of pollutants remains in the environment. The adaptation process is much easier. The environmental agency gradually feels out the right tax by starting out at the bottom of the scale. For a politically strong environmental agency the process has the advantage of permitting the establishment of a tax system even in cases when information about pollution functions is lacking. It allows experimentation with the tax so that the effects on environmental quality of different measures can be observed. Sometimes the tax must be revised. Ideally, possible alterations in the system will be announced in advance, for a particular planning period for example, so that no misallocations occur in the private sector. The disadvantage of this approach is that if the environmental agency is weak, the low charges set in the beginning will remain and the national economy will never achieve target point R.

Strategy III: Equilibrium solution. The environmental agency can aim directly for the optimal solution by setting a charge at point RR′. Businesses and communities can orient their long-term environmental behavior to this charge. The rate must only be revised if the curves shift, as when pollutants are reassessed.

3. A number of problems arise in setting the "correct" tax rate. One difficulty is that the waste products are not identical to pollutants. The wastes discharged into the environment are transported and partly decomposed in the various media. Knowledge of these diffusion models is necessary in order to trace the pollutants to the waste products so that the tax rate can be determined.

4. A number of factors influence the kind and degree of damages. Damages increase as the amounts of pollutants increase. This relationship, as we have already seen in Chapter III, is described as the damage function. The tax must therefore increase when the assimilative capacity of the environment decreases. This can happen quite suddenly and for a relatively short period of time, as during an atmospheric inversion. This requires great flexibility. Since the objective of environmental policy is to maintain a given environmental quality, and not to fix limits for the amounts of pollutants, the policy must aim at reducing the emissions. This is especially true if the environmental medium in question can only absorb a small amount of pollutants. For this reason, taxes must be able to increase so drastically in acute situations that polluters will adapt to the total quantity of tolerable emissions and produce lower levels of pollutants.

Damages also vary with changes in the uses of the environment and with variations in the societal value accorded to those uses. For example, the damage caused by polluted water can increase just by society placing a higher value on drinking water. Or, if the population grows and the demand for drinking water rises, this use of water automatically becomes more important. The tax on water pollution should rise accordingly. The same holds true when the level of ground water falls as a result of increased water runoff caused by construction or by natural changes in the environment.

The tax per unit of pollutant must also increase when the population of an area grows, because the same amount of pollutants then affects more people. This type of increase in societal damage must be reflected in the tax system.

The quantity of pollutants dumped by one business in a region does not in itself determine the tax rate. Rather, it is the value placed on those pollutants in the environment. As we have seen above, this value can change in response to a number of factors on which the business itself has no influence.

One consequence of these considerations is that taxes must constantly be checked against and adapted to new assessments, new information, and new developments. If this is not done, then the tax system itself distorts the production structure and induces misallocations for which unpleasant corrective procedures are necessary. For this reason, a considerable amount of flexibility is needed in order for such a system to function properly. A parliamentary body cannot handle the process. It must be entrusted to an environmental agency with relatively strong authority.

5. A tax or charge system requires the availability of data about the when, where, and who of environmental pollution. Knowledge of a pollutant's origin, its distribution in the environmental media, and its effects on humans and nature is necessary for *every* measure of environmental protection. A prerequisite for solving environmental problems is therefore the development of systematic surveys of data on environmental quality for the various media. The collection of such information is certain to be an expensive process, but the key function of the result in any measure of environmental protection indicates the value that such an investment of time and money has.

6. We do not recommend concentrating policy on individual pollutants or media because it is impossible to establish a single tax for all the possible pollutants in every environmental medium. Uncontrolled pollutants and media would then be automatically exploited. Instead, equivalent taxes must be established for the various media and pollutants. If high taxes are set on air pollution and lower ones on water, substitution processes will develop and lead to an undesirable overloading of water systems. The same phenomenon will occur if pollutants are assigned radically different taxes.

Consider, for example, the problem of thermal pollution caused by nuclear power plants. The production process creates a great deal of heat which can be emitted either into the air or into water. If the government reduces the amount of heat which may be discharged into the air, then managers will switch to water-based cooling processes, and vice versa. It makes economic, if not environmental, sense to take advantage of the uncontrolled medium.

For these reasons, it is clear that we must design a coordinated *system* of taxes which takes into account both the differences and the intricate interdependencies between the media and the pollutants.

7. As the basis for the rate of tax, the assessment of a pollutant's social damage must also include the costs of removing waste products. Even if wastes did not cause health, ecological, or other damages, this economic reason alone would suffice to require the establishment of a charge for each ton of wastes. Above all, the disposal of wastes should not be subsidized. This approach only disguises the societal costs and encourages the production of more wastes.

A relatively high tax on waste products might also generate an incentive to recycle resources. This is the link between waste and resource management. If it becomes obvious that the resources on spaceship earth are becoming scarcer, then a recycling process induced by a tax on wastes promotes the reuse of resources extracted from the environment. The problem can be approached from the other end of the production cycle as well. By taxing the exploitation of raw materials the government encourages multiple use of those materials which have already been extracted.

8. Difficulties arise in measuring damages in physical terms and evaluating them monetarily. If the measurement of damages and its valuation are not possible, the tax may be set according to the standard price approach discussed in Chapter VII. In this case the total quantity of emissions that is tolerable is determined by a political process in which all information available, such as rudimentary knowledge of the damage, should be taken into account.

9. The introduction of new products on the market raises yet other environmental policy problems. It is reported that the chemicals industry alone increases its products by 500 new substances yearly. New products and product variations are an important vehicle of technical progress and economic development. But it should be clear that possible social costs should be included in decisions about developing such products. This implies the need for detailed information about any side effects, which are easily neglected in purely economic considerations of profitability. Most importantly, the new pollutants which might occur as a result of the new products must be identified, their pollution functions

determined, and the amount produced should be taxed, so that the costs are recognized in the economic calculations.

As easy as this might sound in principle, it is incredibly difficult to implement. How can the economic system function if every new product variant has to be administratively approved?

10. Once a tax is set, the problem arises as to the reactions of the environment and the economy. In planning environmental conservation policies, it must be remembered that the effect of any measure is always delayed. For example, Meadows and others have noted that a decrease in the use of DDT in 1970 would first be followed by a severe increase in the DDT content of fish because of the persistence, ubiquity, and accumulation of DDT in food chains. Not until 1995 is it expected to fall back to the 1970 level.[9] Individual economic units also experience time lags in adapting to such new measures as charges. The lack of information is another cause of lags. It can take a long time to identify a problem, to determine the effects of a particular substance on the environment, and then to have the scientific findings recognized on a political level. The political decision-making process can also require a great deal of time, causing yet further delays in the implementation of a policy and its eventual effects.

11. The problems discussed thus far—availability of information about pollutants, equivalency of measures, and lags in effect—arise not only in the setting of taxes, but in the implementation of all other measures of environmental protection as well. One difficulty characteristic of charges and taxes, is the uncertainty about the reaction of business to changes in the price structure. Specifically, it is believed that economic sectors with greater market power and monopolistic market forms will shift the tax to the products. They will then not feel the intended pressure to change their modes of production.

A monopolist is faced with the same adjustment processes as a firm in a competitive market. A constant tax per unit of emission must be paid. The monopolist can reduce his tax burden and thereby raise his profits by abating pollution. Only if one assumes that monopolists do not minimize costs would the emission tax have no effect. However, this assumption is unlikely. If the tax is set at the correct level, it represents an incentive to monopolists, too, to reduce pollutants. The profit maximizing monopolist has the same optimal conditions as a firm in perfect competition. This means, however, that the profit-maximizing monopolist will reduce the same amount of emissions as a firm in a competitive market if the adjustment technology and the level of emissions are comparable. The market form of monopoly therefore does not represent a reason for the malfunctioning of intra-firm adjustment processes. These adaptations are the key reactions to an emission tax. This leads us to the conclusion that the incentive tax will have the same desirable effects in an oligopolistic market structure.[10]

The structure of demand is altered by an emission tax through the principle of forward shifting of costs. The monopolist reduces the supply of goods relative to that provided on a competitive market and raises the price. Assuming that the relevant curves between perfect competition and monopoly are at all comparable, the monopoly price will be higher than that of the good on a competitive market and the supply will be lower after adaptation to the emission tax.

12. Emission taxes not only represent an incentive to abate pollutants with given technology by shifting resources from production to abatement, they also provide an incentive to develop new abatement processes. This is achieved through a decentralized mechanism, namely price signals, rather than through government pressure. It is to be expected that emission standards will stimulate the imagination of businesses in the area of pollution abatement far more effectively than bureaucratic measures such as emission standards.

13. If a substantial rise in the price of a product does not produce a perceptible decrease in consumer demand for the product, then the demand for this good is termed inelastic. Under certain conditions, and at a given income level, the rise in price of the good can lead to a fall in the demand for other products. This has serious implications for environmental policy if the good in question is environmentally degrading. The behavior of demand toward environmentally degrading goods when prices are raised relative to other goods is decisive. A tax or charge system must take into account the structure of demand. An ill-conceived tax can reduce the demand for environmentally sound products in attempting to limit the production and consumption of an environmentally degrading product for which demand is inelastic. This phenomenon affects inputs to production as well as the final market product. If a tax makes a certain necessary input more expensive, businesses will reduce investments in environmentally sound inputs so as to be able to afford the one they need, even if they recognize that it is environmentally degrading.

14. Taxes must be adjusted to changes in income. If income rises with growth and the taxes on environmentally damaging products are not altered accordingly, then the demand for these goods also rises. If, however, the taxes or charges are based on the damages that the product causes, then the tax will rise automatically. For, when economic growth creates greater demand for a degrading good, environmental damages increase. The effect of increasing incomes in the process of growth can therefore be corrected by adapting the taxes to income changes.

The income of individual groups can rise nominally or in real terms as a result of bargaining processes. Charges on environmentally damaging products then have less impact on a group's freedom of action. The effect of the tax, its ability to change the pattern of consumption is thereby significantly weakened. This is particularly true when the taxes trigger new rounds in the distributional struggle and thereby cause a profit or wage-induced inflationary process which permits the consumers to maintain their consumption patterns.

15. Charges and taxes are based on the "polluter pays principle," in contrast to the "victim pays principle" which we discussed above in the context of subsidies. According to the former, the costs of pollution abatement should be assigned to the party that causes the damage. It agrees with our sense of what is right, and it is also more efficient in achieving a socially desirable allocation of production factors. Setting a charge on the pollution guarantees that all the societal costs incurred by production also appear in the economic accounts of the producers, and this approach thereby eliminates the discrepancy between social and private costs.

FOR SALE: THE ENVIRONMENT

Emission taxes have the advantage of providing incentives for the abatement of emissions without taking recourse to direct regulation. Emission norms for firms (or a permit system for new investments) have the advantage of guaranteed reactions. Can a system be devised to provide the benefits of both the emission norms and the tax systems, and at the same time avoid the disadvantages of both approaches? Is there a measure of environmental policy which combines the guaranteed reaction from the private sector and the incentives for further reductions in pollution?

The combination of these two aspects has been proposed by the Canadian economist Dales.[11] We have already seen that the basic cause of environmental pollution is that the private sector uses the environment as a "common property resource" to dispose of waste products without paying for this service. According to Dales, the reason that there is no price on the environment is that no ownership relations or rules have been set for this good. As an institution, private property defines for any item many rights (rights of operation, rights of access) which are socially established and recognized. Property consists of 1) a series of rights for the use of an item, 2) a provision for the inclusion or exclusion of others from use, and 3) a right to sell the item.

Dales distinguishes between four types of property rights. Exclusive property rights cover most rights of disposal, notably the right of sale. This means that a price evolves for this type of property. Even this most extensive form of ownership is regulated by rules, so that the rights of disposal are not arbitary.

A second form of ownership, called status or functional ownership, refers to a series of rights concerning disposal or usage which are accorded to some individuals, and from which others are excluded. Here, the use of an object or a particular service is permitted to certain individuals, and this right is not transferable. The examples cited by Dales include a license to drive a taxi or to notarize documents, admission into a guild, and, in the broadest sense of the term, an import restriction to protect a domestic producer. Such regulatory measures by governments create a series of rights of disposal which usually result in high and secure incomes for some individuals and groups.

A third type of property rights provides access to a public good for a specific purpose. The rules for the use of such goods are clearly defined. An example is the highway system.

Finally, on the other end of the scale, there is the common property resource, which actually represents a "nonproperty." The commons can be used by anyone for any purpose. There are no rules regulating access or use. Examples from this category are the seas, the air, and, to a large extent, inland water bodies.

How do these different types of property rights relate to environmental policy? As we have noted in Chapter IV, it is not possible to parcel off the environment. Even if it were desirable, the environment cannot be privatized in such a way that the use of an environmental service by A will not affect the activities of B. And since no one wants to sell the environment to a corporation that would then resell it for specific purposes, the creation of exclusive property rights for the environment is out of the question. This does not mean, however, and this is the root of Dale's proposal, that the regulations for the use of the commons cannot be changed. The establishment of a tax system which permits the use of the environment only at a price represents such a change. Another alteration is the introduction of emission norms. Yet another possibility in this direction is the sale of pollution rights.

If pollution rights were introduced as a measure of environmental policy, an environmental agency would have to determine a tolerable level of pollution for each period, and then sell pollution permits. Only licensed producers would then be allowed to emit pollutants, and only in the amount specified by the permit. Pollution rights would be auctioned at the beginning of each period (an emissions exchange). Assuming that the total amount of pollution permitted is less than that presently emitted, a price would develop for these permits. The environmental agency could also demand a minimum price at the auction.

The sale of pollution rights involves the same problems as those encountered with other measures of environmental protection. In order to establish the permitted level of pollution and the minimum price for the permit, the environmental agency must determine the present level of pollution. This means that the authorities need information about the damage functions, and they must consider the problems of equivalency between environmental measures for individual pollutants and environmental media. All this is then reflected in the resulting price structure of the permits for the various individual pollutants.

The price of the permit is determined by two factors. The first is the maximum emission level set by the environmental agency, and the second is the demand of industry for the permits. If the agency establishes a large margin for a pollutant, then the relative price of the pollution permit will be low. As a result, industry will tend to substitute inputs and production processes to reduce other pollutants and to take advantage of the least restricted one. The system is able to correct itself to a certain extent in such a case because the increasing demand from industry for permits for this particular pollutant will raise the price, and in the long run scare off other buyers. Although such cases should be avoided by setting the level correctly in the first place, this built-in adjustment mechanism is an advantage not shared by pure emission standards or tax systems.

Permits must be sold on a regional level. If freely marketable permits were sold all over the country they might amass in one area and endanger the environmental quality of the region. Unfortunately, the definition of regions is problematic for several reasons, one being that the rights of pollution must be differentiated according to the environmental media. It cannot be assumed that a ton of pollutant will have the same effects in both air and water. The administrative problem here is that the environmental media have very different spatial dimensions. The boundary determined for air pollution will most probably not coincide with the area defined by the water system. If the agency chooses to set different boundaries for each system, it could provide loopholes for industry, which will be quick to take advantage of another medium in a different region to dispose of pollutants. It is almost impossible for environmental agencies to control or even to foresee all such loopholes.

The introduction of pollution rights could entail another problem which has not yet been discussed. The sale of permits could provide larger businesses with a patent recipe for squeezing smaller firms out of the market. Large industries could use their financial advantage to buy up all the permits in an area, thereby making it impossible for the smaller units to compete. This case is unlikely to occur often in practice, because it assumes that several firms from the same sector are concentrated in a specific region. A greater danger lies in the competition of firms for the regional labor supply. A dominant firm may find it very tempting to buy up pollution permits in order to reduce the demand for labor in the area. This then implies lower wages for the people in the region.

An emissions exchange, a marketable system of pollution rights, represents an ideal solution to the problem of environmental protection *if* the accusation of market distortion can be refuted. The government cannot correct one type of misallocation by inducing new, possibly worse misallocations. Until this problem can be resolved, emission taxes must be considered preferable to pollution rights. It is true that a system of taxes does not provide 100 percent assurance that industry will react, as do emission norms and pollution rights, but it does not distort competition.

THE CONSUMER IS SOVEREIGN

In the last analysis, all production is geared toward the consumer. There is no doubt that every environmental policy must therefore significantly influence consumer behavior. How can this be done? How can consumers be made aware of the impacts of their behavior on the environment, and how can they be induced to alter their patterns of consumption? Besides providing information about the effects on the environment of a given pattern of consumption, there are basically two possible ways of influencing consumers. On the one hand, preferences can be altered, and on the other, consumer options can be reduced.

Attempts to change consumer preferences must focus on developing more environmentally aware behavior among consumers, so that they recognize the environmental effects of their consumption and can alter their buying behavior accordingly. This involves, for example, information campaigns to encourage consumers to buy environmentally sound products and not to buy those which involve high disposal costs. Shaping consumer preferences means creating a new environmental ethic which would give greater weight to environmental than to economic considerations in consumer decisions. The problem with this approach is that it tries to appeal to social interests—classical moral suasion. The questionable success of this tactic has been discussed already. There is relatively little incentive for individuals to change their behavior when it appears that their contribution to the resolution of the problem is merely a drop in the bucket. Any attempt to change consumer preferences must include a stricter control of advertising, especially in those cases where the consumption of environmentally degrading products is being promoted.

With respect to the second approach, there are a variety of ways available to reduce consumer options.

1. Taxation of environmentally damaging products in favor of environmentally sound products[12]: Since the consumer has a set income per period, the increase in price of environmentally damaging products relative to the environmentally sound product should reduce the consumption of the former. The effectiveness of this measure, however, is limited in two cases. If the demand for the environmentally degrading good is inelastic, the imposition of a tax on the good could cause a reduction in the demand of other, possibly more environmentally sound products. Demand patterns must be investigated before introducing a tax of this type. Further, if the introduction of a tax leads to a new round in the distribution struggle, then the limitations on consumer options will not be great enough. The changes in the price structure produced by the tax will be canceled out to a great extent by the real or nominal rises in income, thereby leaving the consumer unaffected.

2. Disposal charges: Until recently, the economic system recognized neither the damage caused by production and consumption patterns, nor the costs of abating the pollution. We have already discussed at some length the possibilities of taxing the emission of pollutants into the environment; the policy maker must also determine ways to cover the costs of disposal. One approach is to tax products which are particularly dangerous or difficult to dispose of, such as nonrecyclable containers or aerosol cans. The tax should then be reflected in the price of the product in order to reduce its attractiveness on the market relative to more environmentally sound products.

3. Definition of product standards: An environmental agency can require manufacturers to change products which have environmentally degrading characteristics, or it can prohibit the production of unsound goods. By thus defining the nature of certain consumer goods, the measure is guaranteed to have an effect on environmental quality.

SMOKER AND NONSMOKER

In many cases it is possible to separate the competing uses of the environment in order to avoid, or at least reduce the environmental problem. Examples are pedestrian malls or smoker and nonsmoker sections in planes and theaters. Activities can be separated according to time, space, or people. A division based on time permits a given activity to be carried out only within a certain period, such as the prohibition of bathing in a reservoir during a drought, or the closing of certain roads to traffic during peak shopping hours. Spatial division of a river, for example, would reserve the water upstream for relatively environmentally sound activities, and would permit environmentally degrading activities downstream only. The separation of competing uses based on persons would limit the number of people using a given environmental service, such as a trail in a national forest. The choice of exclusion mechanisms (prices or a first-come-first-served system) raises questions of equity.

This approach cannot be implemented as easily as the above examples might suggest. A situation in which the complications involved become particularly clear is the spatial separation of residential and industrial areas. This has been suggested in order to reduce the pollution in residential areas caused by industrial production processes. A look at the transportation problems this would entail, however, indicates that the solution is neither economically nor environmentally flawless. New roads would probably have to be constructed, modes of public transportation would have to be devised, and individuals would have to buy cars. People would need more time to get to and from work. These social costs would have to be included in the economic calculation. In addition, the rises in energy consumption for transportation purposes would increase air pollution. Even if the economic and social costs involved were acceptable, the environmental benefits of such a plan would be questionable. The separation between industrial and residential areas remains artificial because diffusion processes within the environmental media still enable the pollutants emitted in the industrial area to spread to the residential area.

The lack of separation between industrial and residential areas is characteristic of our complex urban centers, and is often considered an indication of the inadequate control exerted by the market system. The problem provides ammunition for arguments in support of planning the use of large open spaces. Although the market is to a certain extent responsible for the mingling of industrial and residential areas in urban centers, it is questionable how much better a situation could have been created by planning. Could a city planner seventy or eighty years ago have foreseen the turns that industrial development would take by the 1970s? That would have taken a herculean feat of the imagination. Even if the necessary genius had been at work, is it likely that this person's advice would have been followed? If industries had been located at a sufficient distance from the residential areas, could the population have been expected to travel the long distance to work, especially considering the means of transportation available and the long work hours, just for the benefit of their descendants eighty years later? Can one really be so sure that city planning produces better results for such perspectives than the market mechanism? This question must be posed in order to put the possibilities of a spatial division of competing activities into the proper perspective. Land-use planning is certainly necessary, but the problems it entails are immense.

Another question which must be answered before large-scale environmental planning is attempted, is whether a uniform level of environmental quality is desirable, or whether higher levels of pollution should be permitted in some areas than in others.

Mishan[13] and Dales[14] argue that a uniform level of environmental quality is not a valid goal. The benefits of less polluted areas must be considered, particularly for recreational purposes. If the pollution in region A rises, the value of and the demand for the relatively cleaner environment of B will rise, particularly among the inhabitants of A. The corollary of this thesis is the differentiation between ambient environmental quality and the related differentiation between emission standards. This would mean establishing stricter emission standards in regions with high recreational value. Such a functional division of space must be accompanied by land-use planning on a large scale, and by careful choice of industrial sites.

If independent regional agencies are empowered to establish local environmental standards, then steps must be taken to prevent them from using the issue in competing for industries. Just as local governments try to attract industries to locate in the area by offering tax benefits or promoting investments for infrastructure, they might set lax standards for environmental quality. The choice of a suitable political process for establishing these standards is a problem that cannot be resolved here, but governments must take into account the interests of the affected population and the national goals involved in a functional division of land use.

The regionalization of emission norms corresponds generally to a spatial differentiation of taxes. If the economic policy chooses to use taxes instead of emission norms, how should they be set? Uniform ambient environmental quality standards result in a spatial adaptation of the environment or of the pollutants. A similar reallocation can be achieved through taxes by setting relatively low rates in areas with low levels of pollution, under the assumption that the marginal damage of a unit of pollutant is low in relatively clean areas. In the course of time the pollution in the various areas will level out.

If, on the other hand, a policy of functional differentiation is chosen, then the taxes must be set higher in areas whose environmental purity is particularly valued. Therefore, if more pollutants are permitted in an area, the charges will be lower, and if less pollution is tolerated, the charges will be higher. This relation between the permissible level of pollution and the rate of charge must also be considered when introducing regional pollution permits. If all areas are allocated the same number of permits, then economic activity is certain to move into areas where pollution is low. The end effect is to equalize the level of pollution throughout the country. To avoid this, a functional approach to the regional differentiation of environmental quality should be taken. Fewer permits should be issued (or higher taxes should be levied) in areas whose clean environment should be conserved for special purposes, such as recreational activities. The attractiveness of these areas for industry could be further reduced by setting higher prices on the few permits available.

Large scale land-use planning to provide better environmental quality in residential areas and permit more pollution in industrial centers seems to make good sense. But this approach to environmental control also involves significant problems. How can the residents of industrial centers be expected to tolerate a higher level of pollution than do their neighbors in areas labeled purely residential? Who will establish the permissible level of pollution in each area? Isn't it possible to aim for a uniform level of environmental quality for the inhabitants of all regions in a country without reducing the quality of recreational areas to the level of industrial centers?

PUBLIC INVESTMENT AND OTHER MEASURES

In addition to the key approaches to environmental policy discussed above, certain supplementary measures are available. One possibility is public investment. Here, the government finances investments in the private sector to improve environmental quality.

There are two drawbacks to this measure. First, public investments often do not get to the root of the problem. Investments in equipment to increase the assimilative capacity of the environment by regulating the water level or enriching the oxygen content of the water, for example, do not reduce wastes; at most such measures neutralize for a time the effects of continued pollutive production processes. Often the reverse effect is achieved: industry finds it can create more wastes. This then requires additional investments by the government. There is no incentive for industry to change its mode of production.

The second drawback is that these investments are paid for through taxes. The taxpayer winds up paying for investments which are actually the responsibility of industry. It is true that the citizen will also pay for the introduction of environmentally sound production processes through price rises, but the economic principles involved are different. If new equipment is financed through taxes which are then shifted to the price of the product, then only the consumer of the product in question will pay the price for environmental protection. In the case of public investments, on the other hand, the burden is placed on all taxpayers, regardless of their consumption of the environmentally degrading good.

As we have indicated above, taxes on polluting products cause a restructuring of production, but public investments only veil the problem.

Funds for research on environmental damages and the effects of certain environmental policy measures represent a special type of public investment. Such research is the basis of economic policy decisions, and is necessary because the government cannot rely on information provided by biased sources. The government can also undertake or promote research of recycling and waste disposal processes. Government support of private research into these areas raises the question of the extent to which the results then become public property, and therefore available for all potential uses.

The resolution of environmental problems touches on the important political economic question about the level of government involvement in a national economy and the latitude remaining to the private sector. Environmental policy is a task which will increase government involvement. The nature of this involvement can follow two different courses:

—either the government assumes the responsibility for pollution abatement and finances it through taxes, thereby significantly expanding its involvement;

—or the government sets the conditions for private decisions through such measures as emission taxes, but does not itself undertake to implement environmental protection.

We favor the second alternative, since it does not unnecessarily swell the level of government intervention. It restricts environmental policy to establishing new institutional conditions for private decision making.

CONCLUSION

Alternative environmental policy measures must be evaluated through cost-benefit analyses. The social costs and benefits of measures depend to a large extent on the goal functions and the resulting criteria of society. Enough information is available about the present conditions of the environment and the general values of society to conduct these analyses in terms of the question, "Is measure A better than measure B?"—and no longer "environmental protection: yes or no?"

One important criterion which should be used in judging a measure is its effectiveness. Does it reduce pollution sufficiently? We have seen that emission norms and pollution permits do satisfy this criterion. They guarantee that industry will react rapidly to the policy, and do not involve time-consuming adaption processes.

Other criteria are efficiency (minimization of abatement costs) and the creation of incentives for the reduction of pollution. Does the measure have self-controlling mechanisms, or does an environmental agency have to intervene each time the economic or environmental conditions change? The discussion above has indicated that a tax system is capable of adjusting to new situations to insure a reduction of pollution.

The decentrality or centrality of decision making, the conformity of a measure with the economic system and the organizational and administrative costs are also criteria.

A distinction must be drawn between key measures and supplementary measures. There is a great deal of debate among experts about this issue. The authors believe that only taxes, emission standards, and pollution permits should be considered to be key measures. Subsidies, public investments, the concept of a "basic right" to the environment, and moral suasion are simply insufficient as primary answers to the allocative problem. Inconsequential when used alone, these approaches can aggravate rather than resolve the problem.

Of the major measures available, pollution permits seem to be the most promising, with one basic deficiency. They combine the advantages of a tax system by providing incentives for the reduction of pollution, with the benefits of emission standards by insuring that industry will react rapidly. Unfortunately, pollution permits may severely distort competition in both product and factor markets. We therefore favor emission taxes.

NOTES

1. Cf. Chapter VII, "The Institutional Setting of Environmental Evaluation."
2. N. J. Laudau and P. D. Rheingold, *The Environmental Law Handbook,* New York, 1971, pp. 52 et seq.
3. H. Siebert, "Voerde und eine neue Umweltpolitik," *Wirtschaftsdienst* 58, 1978, pp. 36–40.
4. S. Grundmann and E. Stabenow, op. cit., p. 1783.
5. See Chapter X for a more thorough discussion of this problem.
6. For a more detailed discussion of the problems encountered in implementing an emission tax, see H. Siebert and W. Vogt, *Analyse der Instrumente der Umweltpolitik,* Goettingen, 1976.
7. A. V. Kneese and B. T. Bower, *Managing Water Quality: Economics, Technology, Institutions,* Baltimore, 1968.

8. H. Siebert, *Analyse der Instrumente der Umweltpolitik,* Goettingen, 1976, p. 33.
9. D. Meadows et al., op. cit., p. 70.
10. The formal proof is given in H. Siebert, "Emissionssteuern im Monopol, Eine Anmerkung," *Zeitschrift fuer die Gesamte Staatswissenschaft* 132, 1976, pp. 679–682.
11. J. H. Dales, *Pollution, Property, and Prices: An Essay in Policy Making and Economics,* Toronto, 1968.
12. We have already stated that taxation of pollution intensive goods does not represent an incentive to reduce emissions. On the other hand, since a tax on a product causes a shift in demand, the introduction of an emission tax can be expected to restructure demand and should therefore be applied.
13. E. J. Mishan, *Technology and Growth: The Price We Pay,* New York, 1969.
14. J. H. Dales, op. cit.

Chapter X

International Aspects of Environmental Pollution

A journey of a thousand miles begins with a single step.

Chinese proverb

A special characteristic of environmental pollution is its spatial dimension. The environmental media rarely coincide with man-made boundaries. This has important implications for the development of environmental protection measures.

The spatial dimensions of environmental goods can be classified as follows:

—global environmental goods, such as the earth's atmosphere;

—international environmental goods limited to spatial subsystems of the world, like the Mediterranean Sea and the Gulf of Mexico;

—transfrontier environmental systems which permit the transport of pollutants from one nation to another, for example the Rhine and the acid rains in Sweden;

—national environmental goods;

—regional environmental goods (within one country).

GLOBAL ENVIRONMENTAL GOODS

An environmental system which is used as a public consumer good and as a receptacle for wastes for the earth as a whole is considered a global environmental good. A number of factors complicate attempts to resolve the problem of environmental pollution at this world level. Similar difficulties arise with respect to international environmental goods.

1. In contrast to national environmental problems which can be treated by a national authority with policy measures such as effluent charges or emission standards, there exists no international authority empowered to implement environmental protection measures. Consequently, problems must be solved by international bargaining processes, which involve the divergent interests of two or more states.

2. The lack of information about diffusion processes is a problem of international pollution as well as national pollution. Without this data, it is difficult to determine the source of the pollution and assign responsibility for the damages.

3. The intensity of use differs with each country. Some countries have greater access to the public good (say, beaches), because of varying national preferences, different population densities, and differing levels of per capita income. As a result, the definition of the tolerable level of pollution, the willingness to pay for improved environmental quality, and the bargaining position of the nations may differ considerably.

4. Because environmental services are a public good, it is not possible to exclude from use those nations which are not willing to pay for the conservation of environmental quality. Countries cannot be forced to contribute to financing antipollution measures. The free-rider problem cannot be eliminated in the use of global and international environmental goods.

5. It is possible to "nationalize" parts of international goods, as has been done by some countries in the declaration of 200-mile zones along coastlines. However, this does not provide an answer to environmental problems because international spill-over effects cannot be excluded.

An attempt has been made to control water quality through the establishment of water-management cooperatives at the national level. It is possible that similar cooperatives might be formed to control the quality of the seas.[1] However, the institutional problems which must be overcome to achieve this coordination at the international level are immense. In addition to institutional barriers, it is questionable whether this approach would be of any value to the environment if extended beyond water management to environmental management. While measures of pollution abatement can improve the quality of water in a limited body, it is doubtful whether they would have any effect on the oceans, and even less on such systems as the atmosphere. Once pollutants have entered the atmosphere, compensating measures are of no avail. Such pollution must be controlled at the source of emission.

TRANSFRONTIER POLLUTION

Transfrontier pollution is caused by the transmission of pollutants from one country to another through an environmental system. It must be distinguished from international pollution, which is caused by the emission of pollutants into an environmental medium by at least two nations. Examples of transfrontier pollution are the transmission from France to Germany of potash salts by the Rhine River, of phosphates to Mexico by the Colorado River, and of British and German air pollution to Sweden through acid rains.

The distinction between international and transfrontier pollution is fluid. As a result of varying air and water movements in the environmental system, and different types of access to the public good, one country might be more seriously affected by pollution than others. As it is used here, transfrontier pollution assumes a certain regularity in the diffusion functions. As in the case of international pollution, conflicts of the transfrontier type must be resolved through negotiation between the nations involved, since no relevant supranational organ exists.

Transfrontier pollution can be subdivided into two types: one-way and two-way. One-way transfrontier pollution occurs when the wastes from one country are transported to another, leaving the quality of the environment in the country of origin unaffected. The classic example of this is the pollution carried from a source upstream to a location farther down the river. In two-way transfrontier pollution wastes are also transported back to the country of origin, i.e., through atmospheric conditions and changing winds. The situation becomes more complex when different pollutants are transmitted through different environmental media to the first country from the second.

A variety of approaches have been proposed to control one-way transfrontier pollution:[2]

The establishment of an international environmental agency. Nations could surrender a part of their sovereign rights as regards the environment to an international environmental agency which could tax emissions and thereby control transfrontier ambient environmental quality. The introduction of a tax would create an incentive for the upstream polluter to reduce its emission of pollutants. If countries could agree on an international tax, the national environmental agency could set a supplementary tax on emissions within its own borders. This proposal represents the international application of the polluter-pays principle. However, it is politically unrealistic, since nations are not willing to restrict their sovereignty.

The establishment of pollution rights. In establishing the right to pollute, the polluter and the victim negotiate about who has the right to use the environment as a receptacle for wastes. Clearly, the downstream party has a high stake in persuading the polluting nation to reduce its use of the environment at zero price. On the other hand, the upstream party attempts to continue to emit pollutants without compensating the victim. The result of the bargaining process depends on the relative strength of the parties, which is also affected by international public opinion.[3]

The determination of transfrontier diffusion norms. International diffusion norms could be the focus of discussions between the nations involved or among many countries, such as Western Europe or North America. These norms define the quality of a transnational environmental system when it extends beyond the borders of the polluting nation.

The distribution of the costs of pollution abatement. The costs of pollution abatement could be shared by the countries involved. Here the costs of attaining and maintaining an acceptable level of quality in the transfrontier environmental medium would be added together and distributed among the countries according to a set rate.

The institutionalization of a reciprocal compensation procedure. The mutual compensation principle[4] is based on the assumption that in the bargaining process the polluter will exaggerate the costs of pollution abatement in order to reduce the demands of the other country. Similarly, it is expected that the victim will exaggerate the extent of the damages suffered, in order to maximize the measures to be undertaken. So as to avoid this deliberate falsification of information about the damages and the costs of their abatement, it has been suggested that an international fund be established to which the polluting country would pay according to its assessment of the damages, and the victim according to its assessment of the costs of abatement. This approach is designed to guarantee that the factors determining the emissions tax are set as realistically as possible. The funds collected from the two parties would then be redistributed to them for the implementation of the environmental protection measures. It is important that the countries do not know the rate according to which the money is to be redistributed, because this information would distort their estimates of the costs and damages.

TRADE EFFECTS OF ENVIRONMENTAL POLICY

If the spatial dimensions of an environmental good corresponds to the political boundaries of a country, then the quality of the good can be controlled by national environmental policy. At first sight, it might appear that such environmental goods have no international aspects. This is incorrect. The quality of the environment and its assimilative capacity are significant factors in international trade.

The environment is a factor of production, and is therefore a determining factor of comparative advantage in world trade. As a public good, the environment can influence the comparative price advantage of a particular product. If a country is richly endowed with environmental goods, it will have a trade advantage over nations having few environmental goods. A distinction must be drawn between the following two cases:

a.) A country is richly endowed with certain public goods. To the extent that these goods are nationally restricted (e.g., beauty of landscape), the foreign trade position of the nation in the service sector (such as tourism) is improved. We will not pursue this type of case here.

b.) A country has a higher assimilative capacity than another. Assimilative capacity is determined by four key factors:[5]

—The natural assimilative capacity of the environment.

—The value accorded to the environment as a consumer good.

—The demand for the assimilative services of the environment. This is measured by the amount of emissions released into the environment, which is dependent on the consumption, production, and emissions technology, as well as on the level of development of the national economy. Instead of an assessment of environmental quality, the tolerable level of emissions can be used in the sense of the standard price approach, in other words, environmental policy determines the amount of emissions which exceeds the assimilative capacity of the environment and which the society is willing to accept. This level of tolerable emissions is determined by the preferences, income level, and population density of the society in question.

—The public or private investments undertaken to increase the assimilative capacity of the environment or to decrease the demand for this service.

A basic hypothesis about trade flows is that a nation will export a good if it has a comparative advantage in producing that good. A country can obtain this in three ways:

a.) Through ample supply of the key factors of production needed for the good in question (capital, labor, raw materials);

b.) through favorable productivity brought about by a superiority of technical knowledge (technological, organizational, and management systems, as well as capabilities of the work force); and

c.) through international differences in the conditions of demand.

Policies regulating environmental quality affect the production and sale of a given good. We have already demonstrated, for example, that the price of a polluting good will rise if an emissions tax is established to reduce environmental pollution. This means that the comparative advantage of the country producing this good will be reduced and exports will shrink. In some cases, the price rise will erode the comparative advantage of the nation with respect to this good, so that it would become impossible for the polluting good to be exported.

The Heckscher-Ohlin theorem holds that under internationally identical conditions of demand and technology a country will export the good whose production entails the intensive input of a factor which is abundant in the country. This theorem can be applied to the trade of environment-intensive goods. If one distinguishes between environmentally rich and poor countries, and between pollution-intensive and nonpollution-intensive goods, then one can conclude that the environmentally rich country will export goods whose production is environment-intensive. Environmentally poor countries will export goods whose major input is not the environment. This conclusion assumes, however, that environmental policy correctly reflects environmental abundance.

ENVIRONMENTAL POLICY AND INDUSTRIAL RELOCATION

The comparative price advantage is not only an indicator of competitiveness with respect to international trade, but also provides an indication of the international favorability of a country for industrial location.

Table 8.[6] International Comparison of Industrial Investments for Environmental Protection

Sector		Investments for Environmental Protection						
		Processing industry		Steel and iron	Nonferrous metals	Mineral oil proc.	Chemicals	Pulp & paper
country	total	% of total investment	% of turnover	% of total investment	% of total investment	% of total investment	% of total investment	% of total investment
FRG[a]	DM 8.4 billion	5.2	.28	8.6[e]	8.5	17.4	11.1	10.1
Japan[b]	Y 2385.8[d] billion	14.6[d]	.52[d]	15.8	11.4	26.8	19.8[f]	21.2
USA[c]	$ 16.1 billion	8.3	.37	13.3	21.9	10.5	9.6	19.2

[a]Production and production-related environmental investments for air pollution abatement, water treatment, waste treatment, and noise reduction
[b]Only production-related environmental protection investments
[c]Only production-related environmental protection investments; 1971–1973 only for air pollution abatement and water treatment, 1974–1975 also waste treatment
[d]Processing industry not including pharmaceuticals, ship construction, food and luxuries
[e]Iron-producing industry including drawing plants, cold-rolling plants, and iron, steel, and malleable iron foundries
[f]Chemical and petrochemical industry, not including pharmaceuticals

If environmental policies are enacted in environmentally poor countries, then the production conditions of pollution-intensive sectors will be affected, and the costs of production will rise. This change improves the relative position of an environmentally rich country which does not implement an environmental policy, or at least not as stringently. If capital is internationally mobile, then it is to be expected that, all things remaining equal, caiptal will migrate from the environmentally poor to the environmentally rich country. The relocation of Japanese steel plants in South America is a perfect example of this type of effect.

In the terminology of the debate about qualitative versus quantitative growth, this relocation phenomenon can also be interpreted as a move by Japan to establish a limit on growth in favor of improved environmental quality. This restriction has an effect on all production activities. The comparative cost advantages, the locational advantage of a country thereby disappear. Table 8 provides a survey of investments for environmental protection undertaken in the FRG, Japan, and the United States for the processing industry. The relation of investments for environmental protection to turnover is higher in Japan than in West Germany and the United States. Although the data indicate the magnitude of the problem and some international and sectoral differences, they should be interpreted with care, in view of the weak data base and the different sectoral divisions.

POLLUTE THY NEIGHBOR THROUGH TRADE?

A given country can influence the environmental quality of another in basically two ways: through transfrontier pollution and through trade. The first case has already been discussed at some length above. In the second case, the effects of environmental measures in one country on the production and trade of environmentally damaging products influences the production structure and thereby the environment of another country. If, for example, country A introduces an emissions tax, the price of the taxed good rises, and comparative advantage in trade is reduced. Exports fall, and the production of the pollution-intensive good decreases. Resources are reallocated from the pollution-intensive sector to others and the production in the environmentally sound sector might increase. As a result, the environmental quality in country A should improve. On the other hand, the measure can have negative effects on other countries with which this nation trades. When the comparative advantage of A falls, the comparative disadvantage of the other countries in the trade of the good in question also falls. They might find the production of the good to be remunerative in relation to country A, and therefore increase production. This causes a reallocation of the factors of production in favor of the environmentally degrading good, so emissions increase, and the environmental quality of these countries worsens. In short, the environmental policy of a country affects the quality of the environment in other countries through increased specialization in trade.

Does this pollute-thy-neighbor thesis mean that a country can force a degraded environment on another nation? Can the industrialized nations export their pollution to the developing countries through international trade? Is this a new type of imperialism? The answer to all these questions is no, for the following reasons.

a.) Environmental policy involves costs, both in terms of resources and goal achievement in other areas (full employment, comparative advantage, etc) which limit the country's willingness to implement these measures.

b.) The costs of pollution abatement rise progressively, putting severe limits on environmental policy.

c.) The environmentally rich country can protect itself by introducing environmental policy measures also. By imposing such measures as emission charges on polluting products, it will reduce the attractiveness of these goods for international trade and thereby avoid specializing in the production of environment-intensive products. In this way, the environmentally rich country can maintain or improve the quality of its environment.

Over a period of time, emissions charges will tend to converge internationally if national environmental policies react correctly to changes in the abundance of environmental resources. As we have already seen, a country which is originally endowed with great environmental riches will set a low tax on emissions and will specialize in the production of environment-intensive goods. Environmentally poor countries, on the other hand, will tend to set a high tax on emissions and will specialize in the production of goods which do not place high demands on the environment. The quality of the environment will sink in the former country and rise in the latter. In the long run international trade and specialization will cause emission taxes to converge, if all other things remain equal. A key assumption in this prognosis, however, is that the countries involved have the same level of production technology. If not, the emission taxes are not likely to converge.

ENVIRONMENTAL POLICY AND GAINS FROM TRADE

The pollute-thy-neighbor thesis raises an important problem to which little attention has been devoted as yet. The primary motivation for engaging in foreign trade is the prospect of gains, i.e., the nation hopes to expand its consumption opportunities through foreign trade. If a country specializes in the production and export of environment-intensive goods, the quality of its environment will deteriorate. Up to now, this has not been calculated into the evaluations of gains from trade. Gains should be defined as net improvement in well-being, rather than in terms of pure financial profit. It then becomes necessary to weigh monetary gains against environmental degradation. An open economy must be prepared to accept lower profits for the sake of high environmental quality.[7]

Environmental abundance or scarcity are factors in foreign trade which should be included in calculations of gain like other, traditionally recongized, factors, such as resources, technical know-how, etc. If the value of this factor is not recognized, however, groups with strong interest (export and import industries, unions) might attempt to compensate for the environmental advantages of other countries through trade policy measures. By levying tariffs on imports or subsidizing their own exports environmentally poor countries might attempt to protect the domestic industries that produce environment-intensive goods. For economic reasons they might try to counteract their own environmental policy measures. The costs of such a policy are bound to be high in the long run.

The complex of adjustment processes which occur in environmentally poor countries which implement environmental conservation policies is illustrated in Diagram 15.

ENVIRONMENTAL POLICIES AND DEVELOPING NATIONS

The international division of labor between the industrialized and the developing nations is influenced by the environment's role in determining comparative advantage. It has been suggested that the developing nations have a comparative advantage in the production of pollution-

Diagram 15. Adjustment Process Induced by Environmental Policies

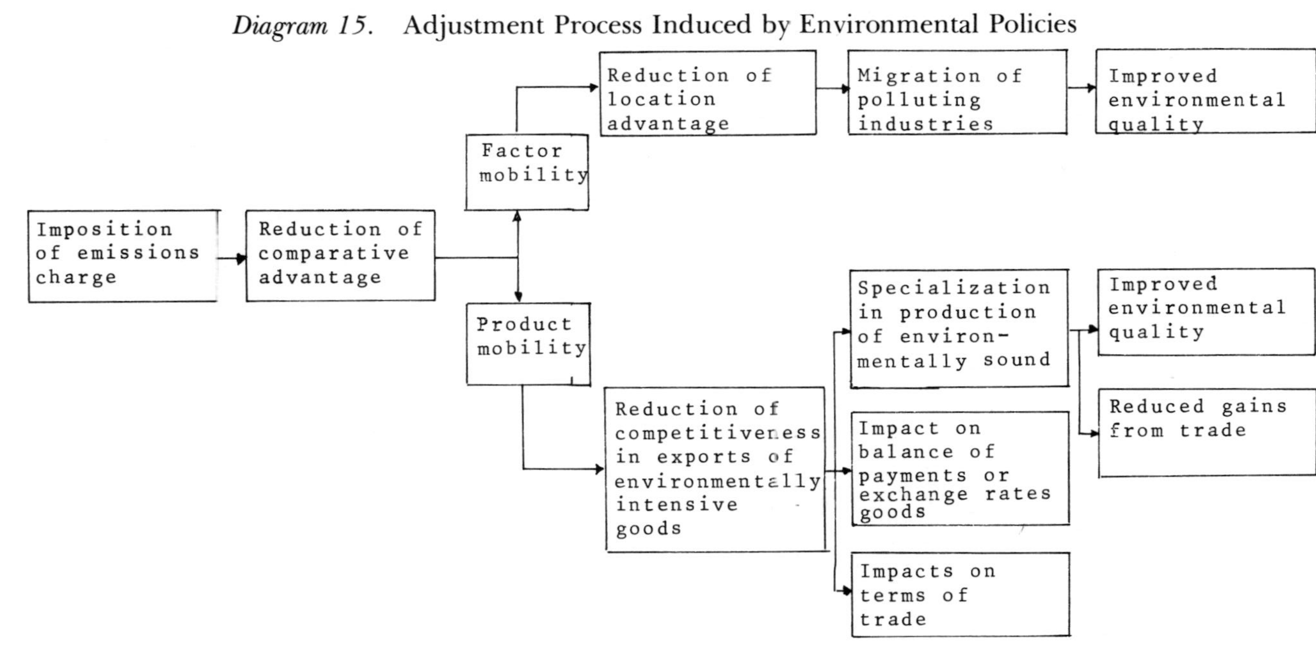

intensive goods. Several reasons have been proposed to support this view. Some developing nations appear to have a high assimilative capacity in relation to the level of the population. Since wastes are a by-product of economic activities and since developing nations have not yet achieved a high level of industrialization, the demand for the assimilative services of the environment is not yet as high in these countries as it is in the industrialized countries. In addition, environmental quality can be accorded different values in developing and in industrialized countries, if assessment is based on income levels or amounts of available private goods.

Another factor contributing to the comparative advantage of developing countries in the production of environmentally degrading goods is that cost and price differences do not vary proportionately with environmental policy measures. Empirical studies show that the costs of waste disposal rise progressively.[8] As a result, the comparative price advantage of the developed nations decreases progressively. Since a comparative price advantage also represents a comparative location advantage, developing nations will attract polluting industries away from industrialized countries and thereby expand their export base.

This type of specialization for the developing countries might appear to be an ideal form of diversification of exports to achieve rapid growth. However, the benefits are only short-term in nature, and serious problems are bound to result from this approach in the long run. An increase in the population, for example, raises the social damage of a given level of pollution; industrial growth multiplies the wastes produced, and thereby also the level of pollution; and, finally, increases in per capita income trigger changes in the value accorded to environmental quality by the population. When these social and economic changes occur, the specialization of the nation in environmentally degrading production must be altered, a process which will incur significant adjustment costs. It is still unclear to what extent future frictions and sectoral adjustments should be considered in present decisions about export diversification.

In reality, the options open to developing nations for export specialization are not independent from the economic policy measures implemented in the developed countries. If, for example, pollution abatement is financed primarily through public investments and subsidies in the developed nation, the exporting industry does not bear the costs, so the price of a polluting product does not rise. The comparative advantage of the industry in the developed nation is not reduced, and it is therefore not ceded to the developing country. When emission taxes or standards are established, the resulting decrease in production of the environmentally degrading good can trigger political pressure for the introduction of export subsidies or import restrictions in order to create so-called "internationally compatible conditions of production." The argument used in the industrialized countries in such a case is that the cheaper imports are not produced under the same environmentally stringent conditions, so import restrictions should be imposed to compensate for the increased costs. These demands disregard that the environment is one of many factors of specialization and that—from the point of view of the optimal international distribution of labor—comparative advantage should not be compensated for by economic policy measures.

Another factor which limits the specialization of developing nations in the production of environment-intensive goods is that the governments of the industrialized countries often do not consider environmental quality important enough to restrict the growth of certain key sectors.[9] Empirical studies have proven that some basic industries, such as energy production, steel, iron, and chemical industries are highly polluting. But since governments have a stake in maintaining a well-rounded industrial structure, they are unwilling to enact measures against environmental pollution which might effectively deal a blow to the industrial base of the economy.

The export potential of the developing nations also depends on the technology in use in the industrialized countries. As noted in Chapter IX, technological development can reduce polluting emissions in the production process, it can provide new methods of waste disposal, introduce resource-conserving production processes, and permit the recycling of material. Such policy measures as emission charges, which cause significant price changes, provide the greatest incentive in the long run for the development of this technology. This technology factor reduces the otherwise expected increase in price differences between developing and industrialized nations. With time the country with sound environmental policies could develop a comparative advantage in the production of investment goods for pollution abatement.

The export options of developing countries are also restricted by the introduction in the industrialized nations of measures to control the pollution generated by the consumption of goods. Such measures may involve standards for agricultural products (limiting levels of DDT), the definition of production inputs (low sulfur content in fuels), or emission norms for technical processes. Since these standards apply to imports as well as to domestic products, the developing countries which are unable to adapt their products will lose their share of the foreign market. The introduction of product standards by a country weakens the export base of its trading partners. Production for the native market is the basis for the expansion of the market into exports, but this base is weakened if products for the native and the export market are subjected to different standards.

This barrier to trade through product standards is aggravated by the introduction of different standards for native and foreign products. This type of protectionism is not permitted by General Agreement on Tariffs and Trade (GATT) regulations. In reality, however, product standards can be formulated in such a way as to permit national industries to meet the standards more easily than foreign competitors. The high-torque engines of foreign cars, for example, have difficulty in adjusting to the lead content of gasoline required by law in the FRG.

Finally, the implementation of the polluter-pays principle in industrialized countries implies that the polluter should also bear the expense of disposing of the unwanted products and materials. The imposition of these costs on the individual firms creates an incentive to recycle materials, thereby reducing the demand of industrial nations for the raw materials of developing countries. This will have an impact on the exports of the developing nations since the demand for raw materials will not grow as rapidly as it might if environmental policy measures were not enacted in the industrialized nations.

If high emission taxes or strict standards raise the price of products from the industrialized countries sold in the developing countries, then import substitutes might become more attractive than the imports from the industrialized world. However, if the demand for the good is not elastic, the terms of trade for the developing countries will worsen.

Environmental policies can also influence the transfer of technology to the developing countries. Emission standards and taxes require the development of environmentally sound technology. When developing nations import such technology they pay high prices because the prices reflect the costs of research and development involved in the production of new abatement technology. Usually this type of technology is not suited to the needs of the developing countries for they are willing to accept production processes which entail higher levels of pollution.

Lastly, as noted in Chapter VIII, developing countries fear that the expenditures of industrialized countries for environmental protection will detract from funds which might otherwise be available for development aid. They resent the policy of richer nations and international organizations which tie strict environmental conditions to credits, particularly when the conditions involve high expenditures.

NOTES

1. A. D. Scott, "The Economics of International Transmission of Pollution," in *OECD, Problems of Environmental Economics,* Paris, 1972; I. Walter, *International Economics of Pollution,* New York, 1975, pp. 154 et seq.

2. Such a solution can be interpreted as an application of the Coase theorem: R. H. Coase, "The Problem of Social Costs," in *Journal of Law and Economics* 3:1–44 (1960).

3. I. Walter, op. cit., pp. 143 et seq.

4. *OECD, The Mutual Compensation Principle, An Economic Instrument for Solving Certain Transfrontier Pollution Problems,* Paris, 1973.

5. H. Siebert, "Trade and Environment," in H. Giersch (ed.), *The International Division of Labor: Problems and Perspectives,* Tuebingen, pp. 108–121 (1974).

6. Source: R. U. Sprenger, "Umweltschutzkosten der deutschen Industrie—Kosteneffekte und Wettbewerbswirksamkeit," *Ifo-Schnelldienst* 8: 1977, Munich, p. 8. The data were developed from information from the Ifo-Institut, the Japan Ministry of International Trade and Industry; Environmental Protection Policy Division; McGraw-Hill Surveys; U.S. Department of Commerce, Bureau of Economic Analysis.

7. Cf. H. Siebert, "Environmental Quality and the Gains from Trade," *Kyklos* 30: pp. 657–673 (1977).

8. GATT Studies in International Trade, Industrial Pollution Control and International Trade, Geneva, 1971, p. 8; see also R. C. d'Arge, "Trade, Environmental Controls, and the Developing Economies," in *Problems of Environmental Economics,* OECD, Paris, 1972.

9. GATT Studies in International Trade, op. cit. p. 13.

Chapter XI

The Political Economy of Environmental Protection: Some Conclusions

We learn from experience that men never learn anything from experience.

George Bernard Shaw

1. Humans need nature. The environment provides our basic consumer goods such as the air we breathe, the water we drink, and the food we eat. It permits our physical regeneration, provides inputs for production processes, and also serves as a receptacle for pollutants. Millions of tons of these pollutants cause severe damage to health, particularly chronic problems; endanger ecological systems whose long-term functions are not yet clear; and reduce the quality of the environment as a public consumer good. Since we have become aware of these interrelationships, environmental quality has started to become an objective of economic policy alongside full employment, price stability, and growth. People recognize that the goal of economic growth can no longer be measured solely according to the increase in goods produced. The measure of social well-being must include environmental quality in addition to the Gross National Product.

2. The economic recession and the energy crisis have overshadowed the concern about the environment. Initiated or planned measures of environmental protection are postponed or down-played because policy makers are afraid that they will put too much pressure on business, or that they will cause price rises, or that they will raise the unemployment rate.

3. The environment issue is here to stay. We can expect increases in

the population, thereby raising the value of the environment and the essentials it provides for life. Further, it is to be expected that private goods will become more abundant in the next few decades, adding to the value accorded to the environment as a public good. This process will occur in the countries of the Third World, too, as they become more developed. The demand for environmental quality will also increase as a result of the projected reduction in working hours and consequent increase in leisure time. These changes, however, will cause increases in consumption and production activities, thereby aggravating environmental stress.

4. Although more attention at present is being paid to problems of resources and energy than to the environment, these questions cannot be separated in the long run. Emission taxes promote recycling and thereby reduce the problem of resource scarcity. Similarly, extraction taxes provide incentives for more economical treatment of natural resources and thus contribute to resolving the environmental problem. The exploitation of natural raw materials, on the other hand, will raise more very serious environmental problems. Strip mining, off-shore drilling, and the extraction of resources under severe climatic conditions in Alaska or eventually Antarctica may create severe environmental disruptions.

5. In the past the environment was used at zero price, permitting the exploitation of environmental resources in any way, for any purpose, as a free good. A zero price does not resolve the problem of competing uses; on the contrary, it aggravates it. Competition is documented in the overuse of a public good whose capacity is limited, or in the existence of conflicting alternative uses of the environment in the production and the consumption sector, as well as between the two. The overuse of environmental goods and services reduces their quality for every potential use.

6. The treatment of the environment as a free good leads to a discrepancy between private and social costs, to a falsification of the control function of prices, to a misallocation of resources, and thereby to a distortion of the production structure in favor of environmentally degrading products. The problem can be resolved in a market economy if certain basic conditions are changed—specifically, the use of the environment as a free good and the discrepancy between private and social costs. However, this requires significant revisions of the market system.

7. We therefore expect the environmental problem to exist in the fu-

ture, and should be aware that new solutions must be designed to extend over several decades. The answer to the problem of the competing uses of the environment does not lie in subsidies, which favor the polluter and are inequitable since they must be financed from general taxes. They lead to no change in the zero price for the environment and therefore maintain the false price structure. The introduction of a basic right to the environment can only constitute a supplementary measure since it, too, does not provide a solution to the competing uses of the environment but only ex post facto damage compensation payments which are difficult to implement. Emission standards as maximum limits for permissible emissions do guarantee a reaction from producers, but they represent serious intervention by government in private production decisions. They do not insure the minimization of abatement costs, nor do they encourage the development of new technologies. They do not indicate how ambient quality norms can be transformed into emission standards for individual firms. Taxes on emissions, on the other hand, provide an incentive to firms to reduce emissions and to develop new abatement technology. The economist is then in a position to indicate the direction in which a solution of the environmental problem lies.

8. The spatial dimension of the environmental conflict raises a series of problems. Nations must decide which institutional measures should be used for the various subsystems of a country, such as regional water-management systems. On the international level, the effects of national environmental policies on trade and industrial relocation must be identified. Finally, the countries of the world are endowed with different stocks of environmental goods. The problems of transnational, international or global environmental systems are particularly difficult to resolve. The pros and cons of these questions will be explored by generations of politicians, lawyers, and economists in papers, textbooks, documents, and international professional conferences. Having hopefully contributed one step, the authors conclude with the reminder of Oscar Wilde's wisdom: "Experience is the name everyone gives to his mistakes."

Bibliography

Anderson, R. J., Crocker, T. D. (1971), "Air Pollution and Residential Property Values," in *Urban Studies* 8:171–180.

d'Arge, R. C. (1971), "Economic Growth and Environmental Quality," in *Swedish Journal of Economics* 73:25–41.

———, Kneese, A. V. (1972), "Environmental Quality and International Trade," in Kay, D. E., and Skolnikoff, E. B. (eds), *World Eco Crisis, International Organization* 26:419–465.

———, Kogiku, K. C. (1973), "Economic Growth and the Natural Environment," in *Review of Economic Studies* 40:61–77.

———, Schultze, W. (1974), "The Coase Proposition, Information Constraints and Long-Run Equilibrium," in *American Economic Review* 63:763–772.

Ayers, R. U., Kneese, A. V. (1969), "Production, Consumption and Externalities," in *American Economic Review* 59:282–297.

Barnett, H. J., Morse, C. (1963), *Scarcity and Growth, The Economics of Natural Resource Availability,* Baltimore.

Bauer, R. (ed) (1972), *Social Indicators,* 2nd ed, Cambridge, Mass.

Baumol, W. J. (1972), "On Taxation and the Control of Externalities," in *American Economic Review* 62:307–322.

———, Oates, W. E. (1971), "The Use of Standards and Prices for Protection of the Environment," in *Swedish Journal of Economics* 73:42–54.

——— (1975), *The Theory of Environmental Policy,* Englewood Cliffs, N.J.: Prentice-Hall.

Bohm, P., (1971), "An Approach to the Problem of Estimating Demand for Public Goods," in *Swedish Journal of Economics* 73:55–66.

———, Kneese, A. V., eds. (1971), *The Economics of the Environment,* London.

Boulding, K. E. (1971), *Economics of Pollution,* New York.

——— (1971), "The Economics of the Coming Spaceship Earth," in Jarret, H. (ed), *Environmental Quality in a Growing Economy,* Baltimore, 1971, pp. 3–15.

Commoner, B. (1973), "The Environmental Costs of Economic Growth," in Schurr, S. H. (ed.), *Energy, Economic Growth, and the Environment,* Baltimore, 1973, pp. 30–66.

Crocker, T. D., and Rogers, A. J. (1971), *Environmental Economics,* Hindsdah.

Dales, J. H., (1968), *Pollution, Property, and Prices,* Toronto.

Dolan, E. G. (1971), *Tanstaafl, The Economic Strategy for Environmental Crisis,* New York.

Dorfman, R., and Dorfman, N. S., eds. (1972), *Economics of the Environment,* New York.

Ehrlich, A. H., and Ehrlich, P. A. (1970), *Population, Resources, Environment,* San Francisco.

Fisher, A. C., Krutilla, I. V., and Cichetti, C. J. (1972), "The Economics of Environmental Preservation: A Theoretical and Empirical Analysis," in *American Economic Review* 62:605–619.

———, Peterson, F. M. (1976), "The Environment in Economics, A Survey," in *Journal of Economic Literature* 14:1–33.

Forsund, F. R. (1975), "The Polluter Pays Principle and Transitional Period Measures in a Dynamic Setting," in *Swedish Journal of Economics* 77:56–68.

Freeman, A. M., Haveman, R. H., and Kneese, A. V. (1973), *The Economics of Environmental Policy,* New York.

Frey, B. S. (1972), *Umweltoekonomie,* Goettingen.

Giersch, H., ed. (1974) *Das Umweltproblem in oekonomischer Sicht,* Tuebingen.

Goldman, M. I. (1972), *The Spoils of Progress: Environmental Pollution in the Soviet Union,* Cambridge, Mass.

Haveman, R. H. (1973), "Common Property, Congestion, and Environmental Pollution," in *Quarterly Journal of Economics* 87:278–287.

Head, J. C. (1974), "Public Policies and Pollution Problems," in *Finanzarchiv, N.F.:* 1–29.

Heller, W. W. (1937), "Coming to Terms with Growth and Environment," in Schurr, S. H. (ed.), *Energy, Economic Growth, and the Environment,* Baltimore, pp. 3–29.

Hite, J. C., and Laurent, E. A. (1972), *Environmental Planning: An Economic Analysis,* New York.

Hotelling, H. (1913), "The Economics of Exhaustible Resources," in *The Journal of Political Economy* 39:137–175.

Institut Economique et Juridique de l'Energie de Grenoble (1975), *Alternatives au nucleaire, Réflexions sur les choix énergétiques de la France* (Rapport préliminarié), Grenoble.

Isard, W., ed. (1972), *Ecologic-Economic Analysis for Regional Development,* New York.

Jarret, H., ed. (1968), *Environmental Quality in a Growing Economy,* Baltimore.

Jevons, W. S. (1865), *The Coal Question,* London.

Kapp, K. W. (1970), "Environmental Disruption and Social Costs, A Challenge to Economics," in *Kyklos 23:*833–848.

Klevarick, A. K., and Kramer, G. H. (1973), "Social Choice on Pollution Management, The Genossenschaften," in *Journal of Political Economy* 2:101–146.

Kneese, A. V. (1971), "Environmental Pollution, Economics and Policy," in *American Economic Review, Papers and Proceedings* 61:153–166.

——— (1971), "Background for the Economic Analysis of Environmental Pollution," in *Swedish Journal of Economics* 73:1–24.

——— (1976), "Natural Resources Policy 1975–1985," in *Journal of Environmental Economics and Management* 3:253–288.

——— (1977), *Economics and the Environment,* Middlesex.

———, Ayres, R. U., and d'Arge, R. C. (1970), *Economics and the Environment: A Materials Balance Approach,* Baltimore.

———, and Bower, B. T., eds. (1972), *Environmental Quality Analysis: Theory and Method in Social Science,* Baltimore.

——— (1968), *Managing Water Quality, Economics, Technology, and Institutions,* Baltimore.

———, and Schultze, C. E. (1975), *Pollution, Price, and Public Policy,* Washington, D.C.

Krutilla, J. V. (1967), "Conservation Reconsidered," in *American Economic Review* 57:777–786.

———, and Fisher, A. C. (1975), *The Economics of Natural Environments, Studies in the Valuation of Commodity and Amenity Resources,* Baltimore.

Lancaster, K. and Lipsey, R. G. (1956–1957), "The General Theory of Second Best," in *Review of Economic Studies* 24:11–32.

Lave, L. B. and Seskin, E. P. (1970), "Air Pollution and Human Health," in *Science* 169:723–733.

Leontief, W. (1970), "Environmental Repercussion and the Economic Structure: An Input-Output Approach," in *Review of Economics and Statistics* 53:262–271.

——— (1973), "National Income, Economic Structure, and Environmental Externalities," in: Moss, M. (ed.), *The Measurement of Economic and Sociology Performance,* New York, pp. 565–579.

——— and Ford, D. (1972), "Air Pollution and the Economic Structure," in Brody, A. and Carter, A. P. (eds.), *Input-Output Techniques,* London, pp. 9–30.

Maler, K. G. (1974), *Environmental Economics: A Theoretical Inquiry,* Baltimore.

Malthus, T. R. (1826), *An Essay on the Principle of Population*, 6th edition, London.

Meadows, D. H., Meadows, D. I., Randers, J., and Behrens, W. W. (1972), *The Limits to Growth,* New York.

Meadows, D. I. and Meadows, D. H., eds. (1973), *Towards Global Equilibrium,* Cambridge, Mass.

Mishan, E. J. (1969), *Growth: The Price We Pay,* London.

——— (1971), "The Post-War Literature on Externalities: An Interpretative Essay," in *Journal of Economic Literature* 91:1–28.

——— (1971), "Pangloss on Pollution," in *The Swedish Journal of Economics* 73:113–120.

Natural Resources Journal, *Coase Theorem Symposium* 13, Part 1:557–716; (1974) Part 2:1–86.

Niehans, J. (1975), "Economic Growth and Decline with Exhaustible Resources," *The Economist* 123:1–22.

Nordhaus, W. D. (1973), "World Dynamics, Measurement Without Data," in *Economic Journal* 83:1156–1183.

OECD (1972), *Problems of Environmental Economics,* Paris.

——— (1974), *The Polluter Pays Principle,* Paris.

Page, T. (1977), *Conservation and Economic Efficiency: An Approach to Materials Policy,* Baltimore.

Pearce, D. W. (1976), *Environmental Economics,* London.

Peskin, H. M. (1972), "National Accounting and the Environment, Oslo.

——— (1975–1976), "A National Accounting Framework for Environmental Assets," in *Journal of Environmental Economics and Management* 2:225–262.

Pethig, R. (1976), "Environmental Aspects of Trade Models," in Walter, I. (ed.), *Studies in International Environmental Economics,* New York, pp. 117–120.

——— (1976), "Pollution, Welfare and Environmental Policy in the Theory of Comparative Advantages," in *Journal of Environmental Economics and Management* 2:160–169.

——— (1977), "International Markets for Secondary Materials, Trade in Secondary Materials, A Theoretical Approach," in Pearce, D. W., and Walter, I., (eds.), *Conference Proceedings of the "Symposium on International and Comparative Economic Dimensions of Waste Recycling and Reuse,* 1977.

———, *Umweltoekonomische Allokation mit Emissionssteuern* (forthcoming).

Pigou, A. C. (1920), *The Economics of Welfare,* 1st edition, London.

Ridker, R. G. (1967), *Economic Costs of Air Pollution,* New York.

——— and Henning, J. A. (1967), "The Determinants of Residential Property Values with Special Reference to Air Pollution," in *Review of Economics and Statistics* 49:246–257.

Rose-Ackerman, S., (1973), "Effluent Charge: A Critique," in *Canadian Journal of Economics* 6:512–528.

Schulze, W. D. (1974), "The Optimal Use of Non-Renewable Resources, The Theory of Extraction," in *Journal of Environmental Economic Management:* 53–73.

Schurr, E., ed. (1971), *Energy, Economic Growth and the Environment,* Washington, D.C.

Siebert, H. (1973), *Das Produzierte Chaos, Oekonomie und Umwelt,* Stuttgart.

——— (1974), "Comparative Advantage and Environmental Policy: A Note," in *Zeitschrift für Nationaloekonomie:* 397–402.

——— (1974), "Environmental Protection and International Specialization," in *Weltwirtschaftliches Archiv* 110:494–508.

——— (1975), "Externalities, Environmental Quality and Allocation," in *Zeitschrift für Wirtschafts- und Sozialwissenschaften* 4:17–32.

——— (in collaboration with W. Vogt) (1976), *Analyse der Instrumente der Umweltpolitik,* Goettingen.

——— (1976), "Environmental Control, Economic Structure and International Trade," in Walter, I., (ed.), *Studies in International Environmental Economics,* New York, 1976, pp. 29–56.

——— (1977), "Die Grundprobleme des Umweltschutzes, Eine wirtschaftstheoretische Analyse," in *Sozialpolitik* 92:141–182.

——— (1977), "Environmental Quality and the Gains from Trade," in *Kyklos* 30:657–673.

——— (1978), *Oekonomische Theorie der Umwelt,* Tuebingen.

——— (1978), "Environmental Policy, Allocation of Resources, Sector Structure, and Comparative Price Advantage," *Zeitschrift für Wirtschafts- und Sozialwissenschaften,* pp. 281–293.

———, I. Walter and K. Zimmermann (1979), eds. *The Regional Dimension of Environmental Policy,* New York University Press, New York.

Solow, R. M., (1964), "The Economics of Resources and the Resources of Economics," in *American Economic Review* 64:1–14.

Tietenberg, T. H. (1973), "Specific Taxes and the Control of Pollution, A General Equilibrium Analysis," in *Quarterly Journal of Economics* 8:503–522.

——— (1973), "Controlling Pollution by Price and Standard Systems: A General Equilibrium Analysis," in *Swedish Journal of Economics* 75:193–203.

——— (1974), "On Taxation and the Control of Externalities: Comment," in *American Economic Review* 64:462–466.

U.S. Council on Environmental Quality, Environmental Quality, *Annual Reports,* Washington, D.C.

Vousden, N. (1973), "Basic Theoretical Issues of Resource Depletion," in *Journal of Economic Theory* 6:126–143.

Walter, I. (1975), *International Economics of Pollution,* London.

——— (ed.) (1976), *Studies in International Environmental Economics,* New York.

Index